MASTERING STRATEGY

The Essentials
with 300 Winning Tactics

CHARLES D. PATTON

Copyright © 2012, 2024 Applied Market Solutions LLC

ISBN: 978-1-963809-48-0

DEDICATION

This book is dedicated to my family, whose roots trace back to Ireland and Scotland and connect with the lineage of General George Patton, my inspiration for this work. I also dedicate it to the few brilliant and strategic individuals, both men and women, with whom I had the privilege to work throughout my career. Notably, the women often stood out for their exceptional intelligence and strategic acumen.

CONTENTS

ACKNOWLEDGMENTS

This book, twelve years in the making, is the result of exhaustive interviews, research into diverse fields, including business and academia, analysis of strategic executive actions, and insights into military tactics applicable to daily life. I have drawn extensively from various sources, each cited in the footnotes, striving to honor the original intent while adapting the content. I am responsible for any inaccuracies, for which I apologize to the misrepresented authors.

I am deeply grateful to those who have supported me in this endeavor: G. H. Bruce for his innovative designs; Denise Katheder and Jane Mishkin for their early encouragement; my brother Dr. Geoff Patton for his motivational push; Carol Gaskin, my patient editor; and especially my wife, Estella Patton, for her immense support and understanding during the long hours spent apart. My heartfelt thanks also go to the countless others who contributed to this project in various ways.

PREFACE

Strategizing involves applying strategy across all aspects of life, enhancing daily decisions and outcomes. Although we may not always recognize it, we use strategies constantly—whether as a company president navigating boardroom challenges, or an entry-level employee vying for a promotion. Strategies pervade various domains: business, sports, politics, personal relationships, and even in the narratives of mystery novels. They are universal, employed by people of every age and background to achieve their desires.

For instance, a child persuading a grandparent to buy an ice cream, or a salesperson competing for shelf space in retail stores, exemplifies strategic thinking in everyday situations. The workplace, however, is a hotspot for strategic maneuvers, both against external competitors and within the organizational hierarchy itself.

This book demystifies strategy, offering insights to help you consistently outmaneuver competition and achieve your goals.

It distinguishes between strategy and tactics—while both are essential for success, they operate at different levels of planning. Strategy shapes the overarching goals and directions, addressing the "why" and "what." In contrast, tactics focus on the "how," detailing the specific actions required to advance strategic goals.

To simplify complex concepts into actionable

knowledge, this book condenses theory and practice into a comprehensive framework. Each chapter builds upon the last, creating a cohesive learning experience that is more effective than the sum of its parts. For maximum benefit, I recommend reading each chapter sequentially.

INTRODUCTION

I still remember the frustration I felt in 1960, when my otherwise excellent high school math teacher struggled to explain the difference between strategy and tactics. It might have been a clear explanation, but as a dyslexic seventeen-year-old, I just couldn't grasp it. I never forgot that moment of confusion, and it became a genesis for choosing to write a book on this topic.

My father claimed that we were related to General George S. Patton, a legendary military strategist. This belief further fueled this curiosity. Although my later genealogical research failed to directly connect our family lines—mainly because many key sources had been destroyed by fire—it was clear that both our families originated from Scotland. Inspired by my father's tales, this perceived connection motivated me and my siblings to believe we could naturally think and act strategically.

Years later, these thoughts led me to deeply explore strategy and tactics. Having gathered extensive information, I am now eager to share these insights with others who, like me, seek to master strategic thinking in daily life.

In the upcoming chapters, I will guide you through the fundamental aspects of strategy, illustrating how these concepts can be woven into a robust framework for practical use. Imagine these strategies as threads in a fabric you will craft into a garment tailored for maximum impact in your life. You will encounter a variety of offensive and defensive strategies that are applicable to numerous scenarios. Consider these as "thought starters" and not a totally definitive compendium. To keep the focus on practical application, I have intentionally omitted complex scientific theories on strategy.

Can strategy be learned, or is it an innate talent? I believe that if you can read, remember, and think, you can master strategic thinking. This book will not only help you understand the power and structure of strategy but also show you how to apply it effectively. You'll gain a critical, yet often misunderstood, life skill. Keep this book in your library as a reference to continually hone your strategic skills and find inspiration whenever you face new challenges.

PART I

Fundamentals of Strategy

What Is Strategy?

After hearing from some famous strategists and observing the structure that follows, you will realize why no single, simple definition can fully explain this complex subject.

Lou Holtz, a legendary football coach, exemplified strategic mastery through meticulous preparation long before his iconic tenure at Notre Dame and USC. In a notable 1989 speech at the University of Notre Dame, Holtz shared strategic principles derived from Sun Tzu, the ancient Chinese military strategist:

- Only take into battle what you need.

- Treat your team members as you would your own children; they will follow you into the direst circumstances.

- Never retreat.

Similarly, Emil Schalk, reflecting on the U.S. Civil War in 1863, emphasized the importance of speed in executing plans: "As soon as you have made your plan and decided to act, move swiftly to achieve your objective before your opponent can react."

General George S. Patton concisely captured the essence of strategy with his advice, "A good plan today is better than a perfect plan tomorrow." From Patton's perspective, our initial working definition of strategy becomes clear: **Strategy starts with a good plan**.

From my research and the study of recognized strategic leaders, an understanding of strategy and its many facets began to emerge—resources (warriors, money), plans, objectives, control, speed, bravery, morale, and more. From these observations, as shown the next chapter, a framework for the mastery of strategy also emerges.

As for tactics, it should become clear later that tactics are mainly small strategies, the parts of a strategy, changes in a strategy necessitated during execution, or steps in implementing a strategy.

Critical Success Factors

A few years ago, an astute and shrewd coworker, Larry D., shared a story about how he managed a particularly intrusive salesperson. Each time this salesperson visited, he attempted to read documents on Larry's desk upside down and unsuccessfully tried to extract information about competitors' bids.

One day, anticipating the salesperson's visit, Larry planned a clever ruse. He arranged for his assistant to call him out of the office five minutes into their meeting. Before leaving, Larry prepared a decoy: he took a competitor's bid, obscured the dollar amounts with White-Out®, and taped the document to a soda cup filled with BBs. He then placed it face down on his desk, ensuring the document was angled so the salesperson could see most of it. The competitor's name was visible, but the BBs concealed the critical financial details.

As planned, when the salesperson arrived and their meeting began, Larry's assistant called him out of his office on schedule. Three minutes later, upon returning, Larry caught the salesperson on his knees, frantically scrambling to collect the BBs scattered across the floor.

Clear Motive: To get even with the salesperson for unethical behavior.

1. **Clear Objective**: To teach the salesperson a lesson and change his behavior for life.
2. **Controlled Environment**: His own office, where he had full control over the situation.
3. **Power**: Access to necessary resources—his assistant, the BBs, and the cup.
4. **Planned Sequence**: A pre-determined plan with precise control over timing, starting five minutes into the meeting.
5. **Perfect Execution**: Flawless in timing and nerve.
6. **Carefully Selected Offense**: Employing distraction and surprise to gain an advantage.
7. **Proactive Defense**: The best defense is a good offense—ensuring no consequences if the cup remained unraised.
8. **Managing Consequences**: Choosing to laugh with the salesperson, not at him, to showcase the victory in a light-hearted manner.

Of course, Larry's plan may not have accounted for **all possible outcomes**. For instance, his assistant might have forgotten to call him out at the designated time, or the salesperson could have reacted aggressively to the embarrassment. Additionally, the salesperson might not have even looked at the invoice, turning the joke back on Larry. However, the joke did land on the salesperson, and everyone, including him, ended up sharing a hearty laugh. Now, let's explore these critical factors for a successful strategy.

1. Motives, Attitude, and Morale

The relationship between motives, attitude, and morale are foundational elements underpinning effective strategic implementation.

Motives

Your motive is your intention—the reason behind your actions. Understanding motives is crucial to grasping strategies and recognizing when they are in play. Your motive is the underlying reason if you are working toward an objective. Motives can range from altruistic to malicious and may be apparent or concealed. Every conflict begins with a motive, and its resolution or failure to resolve marks the end of that conflict. Consequently, motives are pivotal in resolving any conflict.

Motives and their underlying intentions are crucial for understanding your behavior and your opponent's. By grasping your own motives, you can sharpen your focus on your most important objectives and anticipate potential tactics against you. Similarly, understanding your opponent's motives can help you predict their actions and devise strategies to turn these insights to your advantage.

Your explanation captures the essence of how motives can vary and influence behavior. Here's a minor refinement to enhance clarity and readability:

Motives are often concealed, yet you might discern or deduce them from your opponent's actions. They are driven by emotions such as fear, hate, anger, pride, jealousy, lust, a quest for power, or envy. However, motives can also be noble, driven by generosity, humility, enthusiasm, and compassion. Here are examples of common motives:

- **Fear**: Arises from the instinctual reaction to perceived imminent danger, negative judgment, or risks that often feel uncontrollable or incomprehensible. It typically stems from a lack of knowledge, experience, or preparation for a particular situation. Fear can provoke defensive or preemptive actions.

- **Power**: The desire to control or dominate others, often rooted in one's insecurities or ambitions. This need may drive individuals to accumulate wealth, influence, and territories. The quest for power can originate from fears such as poverty and evolve into a relentless pursuit of authority for its own sake.

- **Pride**: Involves defending one's principles or ego. It can motivate both defensive and offensive actions, similar to fear. On a national level, pride can either drive positive outcomes or destructive actions, as seen in historical leaders who have mobilized nations towards aggression or defense.

- **Discontent**: Arises from a compelling desire to alter the status quo, often due to oppressive, unfair, or abusive conditions. While some fear change, others embrace it, and a few actively seek to create it. Those driven by discontent are often catalysts for change, targeting current leadership as the root of societal issues, even when external circumstances may be the true cause.

- **Revenge**: Often fueled by a sense of injustice or a desire to right a perceived wrong, revenge is a powerful motivator that can drive individuals and groups to act in retaliation. It typically has a unique emotional component related to personal vendetta or historical grievances.

- **Ambition**: The desire to achieve a certain status, power, or success, which can motivate expansive and aggressive strategies. Ambition drives individuals to set high goals and persist in overcoming obstacles, often inspiring significant organizational or personal change.

- **Curiosity**: The drive to explore, discover, and understand, which can lead to innovative strategies in business and technology. Curiosity motivates the pursuit of new knowledge and can lead to breakthroughs that disrupt traditional practices and markets.

- **Fear of Missing Out (FOMO)**: This anxiety that an exciting or interesting event may currently be happening elsewhere can lead to strategies designed to quickly capitalize on opportunities before they disappear, often resulting in hasty or impulsive decisions.

- **Altruism**: A selfless concern for the well-being of others, which can drive strategies focused on humanitarian efforts and corporate social responsibility. Altruistic strategies often prioritize collective benefit over personal gain, influencing decisions in areas like environmental conservation, social justice, and community support.

Motives, such as those outlined above, are dynamic—they expand and contract as objectives are achieved and emotions evolve during conflict. You must stay vigilant to any shifts in motives, as these changes can reveal vulnerabilities or newfound strengths in your opponent or yourself.

Motives and strategies rely fundamentally on people. People can execute tasks correctly or make mistakes. They can adapt to setbacks or crumble under success. People can succumb to fatigue, breakdowns, or even death. Yet, they are also capable of remarkable resilience and can achieve greatness under the right conditions. Since your success hinges on your team, choose your people wisely, train them well, and treat them respectfully.

Understanding everyone's motives is critical when developing a strategy. For instance, individuals who have experienced poverty and later attained wealth often harbor a deep fear of losing their assets. Criminals may think intimidating just one juror could result in a hung jury. Government officials have historically been vulnerable to corruption when their desire for financial gain overshadows their integrity. By thoroughly knowing your opponent, you might uncover exploitable weaknesses.

The intensity and direction of an opponent's strategy can often be predicted by understanding their motives. A classic example is from the movie Patton, where the general famously declares, "Rommel, you magnificent bastard, I read your book!" This line illustrates Patton's deep understanding of his adversary's strategies and motives.

Opponents are complex individuals whose predictability may vary. One method to gauge an opponent's motives is to examine their lifestyle choices. What image do they strive to project? Are they emulating a role model, a childhood hero, or perhaps a parent? Their lifestyle often mirrors their needs and their definition of success.

By knowing an opponent's motives, you can discern their targeted objectives, tolerance for change, and commitment to their goals. An opponent willing to sacrifice comfort and even life for a cause presents a formidable challenge compared to one who avoids discomfort and risk. Commitment is often the deciding factor between success and failure in strategic endeavors.

Understanding your motives is equally crucial, as your opponent may exploit them. If money motivates you, options include seeking a better job, enhancing your education, taking a second job, acquiring a company, living frugally, or even committing a crime. The chosen path should reflect your risk tolerance and resource availability. Remember, betting heavily against great odds usually leads to loss. It's essential to manage risks wisely and not gamble beyond your means.

Motives are pivotal for predicting behavior, and understanding your and your opponent's motives is the first step in crafting a successful strategy. Recall the iconic scene in Patton, where George C. Scott, portraying General Patton, boldly shoots at an enemy plane with his pistol. More than his actions, Patton's fearless attitude left a lasting impression. His motive was to demonstrate unwavering courage to inspire his officers and troops.

Attitudes

Attitudes are critical in fostering a high level of commitment. Cultivating a superior attitude without slipping into arrogance is essential—lacking confidence or competence can jeopardize your effectiveness. Strong attitudes lead to strong commitments, while weak attitudes result in weak commitments and can cause loss of focus and failure.

As a key part of your strategy, it's essential to factor in the influence of attitudes and develop effective ways to manage them. The attitudes of yourself, your team, your opponents, and their supporters are all pivotal to your success. You can witness the immediate impact of defeat on attitudes by observing the losing tribe in the reality TV show Survivor; their morale visibly wanes after a failed challenge or immunity.

Positive attitudes stem from being well-disciplined, well-practiced, well-planned, and well-organized. Conversely, poor attitudes often arise from fatigue, hunger, or financial constraints. They can also develop from a lack of commitment, exclusion from crucial information and decisions, or feeling unappreciated. Such negative attitudes can spiral, leading to even worse morale and, ultimately, failure.

Failure tends to deteriorate attitudes further, even temporarily, so it is crucial for you as a leader to actively counteract the ensuing discouragement after a defeat or setback. Employ tact and, when necessary, assertiveness to formulate a new plan, reorganize, and re-engage your resources.

Maintaining a positive attitude is crucial in any strategic endeavor, and one key component of this is perseverance. Perseverance involves steadfastness in doing something despite difficulty or delay in achieving success. It requires resisting pressures from several sources:

- **Fatigue**: Physical exhaustion or lack of sleep can lead to errors. It's vital to manage energy levels and ensure adequate rest.

- **Unsubstantiated Reports**: Do not let threats, rumors, or exaggerations trigger knee-jerk reactions. Always seek additional information before making decisions.

- **Hastiness**: Rushing through tasks can result in mistakes and poor outcomes. Slowing down can help to ensure more accurate and effective actions.

- **Negligence**: Carelessness can exacerbate existing problems. It's important to maintain vigilance and attention to detail in all endeavors.

- **Misguided Sense of Duty**: Avoid stubborn persistence based on a misguided sense of obligation. Evaluate situations objectively to know when persistence is warranted and when it is not.

- **Obstacles**: Challenges, whether natural or human-made, are inevitable. Overcoming these barriers is a testament to your strategic perseverance.

- **Disloyalty**: Betrayal by those you trust can be a severe setback. However, as Tom Clancy wrote in The Teeth of the Tiger, "Treason is only possible from those whom you trust." Learn to move past these events constructively.

- **Dislike by Others**: Leadership and strategic success aren't always about popularity. Winning may sometimes mean making tough decisions that aren't universally liked.

- **Emotional Strains**: Stress and emotional turmoil can cloud judgment. It's crucial to develop emotional resilience and maintain focus on strategic goals.

- **Complacency**: Avoid becoming too comfortable with past successes. Continuous improvement and vigilance are necessary to sustain momentum and adapt to changing circumstances.

By effectively addressing these pressures, you can maintain a positive attitude and enhance your ability to persevere through challenges, driving toward successful outcomes in both personal and professional settings.

Another critical aspect of attitude is the appropriate management of boldness. In his 1832 seminal work On War, Carl von Clausewitz advises against being "foolhardy without object, reckless without caution, headstrong without planning, or disobedient against experienced judgment." Embrace boldness, daring, and bravery, but always with consideration for others' safety, thoughtful planning, and emotional discipline.

Your opponent's attitude is equally critical to your success. It is essential to ensure that your attitude surpasses your opponent's. You can exploit your opponent's fears, present exaggerated facts about your capabilities, and train more rigorously. Alternatively, if your opponent tends to overreact, you might calm their worries, downplay your strengths, and secretly enhance your capabilities, surprising them with your preparedness when they commit to an attack.

Attitudes among allies can also vary, particularly regarding risk tolerance. Each ally has its motives and interests, which can complicate joint efforts. During a critical offensive, divergent risk appetites may strain alliances. As Thomas C. Schelling notes in The Strategy of Conflict, the true measure of an ally's reliability is observed in their actions rather than their words—actions reveal truth. This underscores the importance of trust and mutual understanding in teamwork scenarios.

Managing team attitudes requires aligning individual beliefs, feelings, and desires to foster cohesive morale. If you work solo, maintain strong confidence, a positive outlook, and a clear focus to navigate challenges effectively.

Morale

Morale acts as the barometer of attitude. It is a dynamic element that you can influence within yourself and your team and among your opponents, spanning all levels of command. For instance, when Japan attacked Pearl Harbor, contrary to their intentions of demoralizing Americans, they inadvertently galvanized them into mounting a massive retaliatory effort. Similarly, Bobby Knight's infamous tactic of throwing chairs to motivate his players sometimes spurred them to victory, yet at other times, it completely backfired.

Managing morale is a complex task, but history's most resilient leaders have left us with valuable lessons. The diaries of mid-nineteenth-century Arctic explorers outline five key steps for boosting morale:

1. **Set Clear Goals**: Establish precise objectives that everyone understands.

2. **Develop a Plan**: Outline a strategic approach to achieve these goals.

3. **Encourage Participation**: Foster a collaborative environment where team members to agree, suggest modifications, or

express disagreements with the plan.

4. **Support Divergent Paths**: Respectfully allow dissenters to depart, providing necessary support and keeping communication open for future collaboration.

5. **Mobilize Commitment**: Engage and empower those committed to the plan to begin immediate implementation.

By adhering to these steps, leaders sustained high morale even during prolonged periods of hardship, such as starvation and deprivation. Their esprit de corps stemmed from the leader's integrity, ability to inspire confidence, prioritization of team welfare, and focus on the collective goal. If dissenters cannot be easily persuaded to leave, consider reassigning them to "special assignments" requiring intense innovation and effort or minimal involvement. This strategy ensures they contribute positively or do not interfere with the team's progress.

Motivating people often involves sparking their initiative. While some are self-motivated—a rare trait—most respond to external stimuli. Ironically, external motivation must trigger internal drive for it to be effective. Creating the right environment can lead individuals to motivate themselves. Sometimes, simply capturing their attention and providing logical reasoning suffices, while other times, appealing to their emotions or imposing tight deadlines is necessary. Promoting high morale can also be achieved through fostering teamwork and striving for excellence in various domains, such as speed, resilience, diligence, courtesy, thoroughness, cost-efficiency, profitability, or creativity. Celebrating successes, a key aspect of maintaining high morale, not only acknowledges the team's achievements but also motivates them to strive for more. Encourage promotions, and cultivate a positive environment with mottos and themes, further boosting team morale.

In urgent scenarios, direct orders may be necessary, especially when there is no time for consultation or when decisive action benefits the organization as a whole. However, frequent use of direct commands should be avoided as it can breed resentment and place all responsibility on the leader. A more inclusive approach, involving the team in decisions, not only leads to a more forgiving atmosphere, even in the face of setbacks, but also makes the team feel valued and included in the decision-making process, enhancing their commitment to the team's goals.

Motives, attitudes, and morale are pivotal elements of any strategy, intricately linked to why and how you execute your actions. They shape the effectiveness of your strategy and precede the 'what'—the specific actions you undertake, which are deeply influenced by your underlying reasons.

2. Objectives

Motives typically drive you toward a primary goal, such as earning a promotion, winning someone's affection, conquering a region, or retiring early. This main objective is your ultimate measure of success, but it's also the culmination of several intermediate objectives. These intermediate objectives serve distinct purposes: they help you acquire necessary resources and improve control over the sequence of events, each step bringing you closer to your final goal.

Selecting the right intermediate objectives is critical in strategic planning. Making the wrong choice can waste resources, increase risks, and consume valuable time. While one or two missteps in intermediate objectives might not derail your ultimate goal, accumulating too many can prevent you from reaching it. Additionally, achieving an intermediate goal may lead to complacency, potentially causing you to alter or abandon your original intent.

A common pitfall in failed strategies is mistaking a mere list of objectives for a strategy. Consider a billion-dollar company that outlined a supposed strategy through a list that included generic actions like training staff, researching new markets, and increasing sales. This list lacks strategic coherence—it shows no logical sequence, offers no milestones for measurement, and contains inherent conflicts. A more effective set of objectives might specify training ten staff members in Microsoft

Excel, researching the lightweight laptop market in New York, and aiming to increase sales by $2.4 million by launching a new lightweight laptop. These objectives are specific, measurable, and achievable, forming a robust strategy.

Anticipating potential obstacles and understanding your opponent's strategies are crucial elements of successful planning. To effectively counter your opponent, consider these critical questions:

1. **Opponent's Main Objective**: What is the ultimate goal your opponent is striving to achieve?

2. **Intermediate Objectives**: What steps must your opponent take to achieve their primary objective?

3. **Resource Requirements**: What resources are essential for your opponent, and are they readily available or achievable through intermediate goals?

4. **Critical Intermediate Objectives**: Which steps are vital for your opponent's progress?

5. **Optimal Points of Interference**: Where can you most effectively hinder your opponent's progress?

6. **Alternative Strategies**: If your opponent fails at any point, what alternate paths might they pursue?

7. **Potential Alliances**: Who could ally with your opponent, and how can you disrupt these alliances?

8. **Distraction Tactics**: Which of your opponent's objectives could you target to divert their focus and resources, allowing you time to strengthen your defenses and advance your own goals?

Selecting the right ones is the most critical success factor for any primary or intermediate objective. Equally important is identifying which of your intermediate objectives your opponent might target. You must defend these crucial points without losing momentum toward your subsequent intermediate objectives and your main goal—a delicate balancing act.

Once you understand why, how, and what you aim to achieve, it's essential to consider the "game board"—the field of action where these strategies will unfold.

Fights are not just in War

War is not the only arena for fights and failures. In business, Coke® and Pepsi® slug it out every day. Coke® failed badly in introducing the New Coke®. Pepsi® surely has had its product introduction failures too. They both expanded by acquiring companies selling other beverages, such as Snapple®, orange juice, Gatorade®, and others. Wang had total dominance of the word processing business until the personal computer became popular, and they were slow to pivot to the advancement, leaving the door open for Microsoft® to walk in and take the business with its Word® product.

Politicians wage battle every year with more than half losing. Politicians also wage war and can win or fail. They plan their campaigns, attack opponents personally or on the issues, raise money, gauge timing, plan communications, and allocate resources. Business, politics, war, and human endeavors all wage daily battles.

3. Environment

You operate within multiple complex environments—geographical, political, physical, and psychological—dynamic and ever-changing. Some of these environments also have specific rules. Together, they define your "playing field." To succeed in your strategy, you must be thoroughly familiar with this playing field—the more familiar, the better.

For instance, if you are scheduled to speak in front of a large audience for the first time, visit the venue beforehand. Ensure that the equipment functions correctly, that backup parts like projector bulbs are available, and that you are comfortable with the surroundings. By doing so, you eliminate several variables that could contribute to nervousness.

If you have an opponent, study their past behavior and find ways to use it to your advantage. Opponents often act in predictable patterns. Understanding these patterns allows you to anticipate their actions, reactions, and temperament. Consider how a mob assassin operates: they study their target's behavior, noting where they live, travel, and eat. When a predictable pattern emerges, the assassin plans the attack around this predictability, selecting the safest place and the best time for the strike. While this example may seem extreme, the lesson is clear: avoid being predictable and understand your opponent's predictable behaviors.

Geography

Your strategy must account for various factors such as terrain, weather, distances, daylight and darkness, altitude, and more. It's crucial to anticipate both your view and your opponent's perspective of the playing field.

- **Terrain**: describes the features of the local area that may impede or assist the movement of your resources. It can create natural barriers, set lines of division, and offer disguising cover. Typically, the defender is more familiar with the terrain than the attacker. When planning, consider that terrain can be an obstacle, limit visibility, provide protection, or slow progress due to unfamiliarity. Think about the first time you did your annual tax return. Your lack of familiarity with the process, rules, and paperwork likely made you cautious and forced you to take your time. The following year, with familiarity, the process went more smoothly. Similarly, starting a new job involves learning who controls what, who is in charge, and who can help or hinder your progress. Terrain, in this context, encompasses the rules, expectations, and layout of any environment.
- **Weather**: is another natural factor, only partially predictable. While it affects you and your opponent, being better prepared can give you an advantage. For instance, D-Day was delayed due to weather conditions. Similarly, don't just hope for good weather when

planning a company picnic—have a backup plan for rain.

- **Friction**: refers to the wear and tear on your resources, which can occur in rugged terrain. It affects equipment and people alike, manifesting as delayed flights leading to frayed nerves or missed appointments. Work friction can stem from personality conflicts, internal competition, unclear job duties, and inexperienced coworkers. Friction takes a toll on resources, so it's essential to eliminate it whenever possible.

Political

Your strategy must navigate laws, hidden and vested interests, allies, and conspiracies. War is inherently political, and most personal and business conflicts involve struggles for power and control.

Physical Limitations

People and equipment have capacities and limitations. Individuals, animals, and vehicles can only carry so much load over a certain distance within a given period. For example, mules will sit down if overloaded, and people and machines require food or fuel to continue functioning. Periodic rest and repairs are essential. Supply lines have limited capacities depending on distance and time. Some people and equipment are more efficient for specific tasks than others. Understanding the limitations of your resources and those of your opponent is crucial.

Differences in physical capabilities often decide crucial battles. During World War II, General Patton's tanks ran out of fuel while reinforcing American forces, having outrun their supply lines. This delay could have prolonged the war and led to a major defeat. Similarly, in business, rushing to fulfill a critical order can result in mistakes just as damaging as late delivery.

Psychological and Physiological Condition

Stress, mental fatigue, fear, lack of confidence, overconfidence, group dynamics, and other psychological and physiological conditions can determine your success or failure. Attitudes can exacerbate or mitigate these conditions.

Physical strength alone is not enough; mental and emotional preparedness are crucial. The psychological state of your team can impact how many resources you need to achieve your objectives and affect their overall effectiveness. For instance, harsher environments reduce resource effectiveness, necessitating more resources and support to maintain mental and physical fitness. Environmental factors can also affect the time required to reach your objectives. More time will be necessary if you need additional resources due to mental or physical limitations.

You can minimize the impact of environmental factors on your strategy by ensuring you have all the necessary resources in place and by being well-prepared, thoroughly trained, and extensively practiced in the harshest conditions.

Change

Change is the only constant in the environment. Today differs from the "good old days," and tomorrow will differ from today. You can adapt to change and thrive or resist it and become obsolete. Embracing change means staying alert to new developments and even driving change yourself. Those in power and those with persistence are often the catalysts for change. It's essential to anticipate and prepare for potential changes as much as possible.

Rules

Understanding the rules is essential in every strategy, whether they are the rules of the game, the road, the law, the Geneva Convention, or any other relevant regulations.

Rules apply in all life situations, even in dictatorships. They set the standards of behavior that must be followed and are usually measurable, often including sanctions for non-compliance. Rules encompass instructions, laws, ethics, regulations, practices, religious principles, standards, policies, and precedents.

Typically, the strongest powers create and enforce the rules. Enforcement can range from the brutal methods of a dictatorship to the subtle pressures of religious guilt. If rules are consistently enforced, knowing them better than your opponent gives you an advantage.

Understanding the consequences of breaking the rules helps you decide when it might be worth it. Sometimes, strategic offenses require bending or breaking rules to wrest control from powerful rule-makers. You must weigh the penalties against the benefits. For instance, committing "pass interference" in American football can be worth the penalty if it prevents a touchdown. Be prepared for the possibility that rules may change or that you might not know all the rules.

Elders typically know the rules best, having learned from years of experience. They exert influence through their deep understanding and ability to enforce rewards and punishments. This behavior is typical in government and corporate environments, where senior executives navigate complex rules, exchanges of favors, and relationship networks.

Rules exist to control behavior, but they can also change. Laws vary by government, ethics by culture, and regulations by circumstances. What is ethical in one context may not be in another.

Ultimately, your behavior defines your trustworthiness and integrity—predictability and consistency are virtues among friends but liabilities among enemies who can exploit them. Violating rules can stem from motives like money, ego, lack of information, or ignorance.

Assuming you value your integrity, the essential tests for any decision are: Is it legal? Would independent judges consider it fair? Does it harm anyone? Have you been honest with those affected?

Can your conscience live with it? In conflicts, your opponent may operate with different rules, so be prepared for this reality. Further on in this book, you will read a perfect example of this between United Airlines and American Airlines.

4. Power

Power derives from resources, force, strength, concentration, and information; these elements are intertwined and form the core of strategy. Additionally, power involves allies, reserves, balance, human nature, and the proper selection and application of resources.

Resources

Your resources—money, people, fuel, food, water, materials, equipment, and machinery—largely determine your power. These resources provide the energy to drive your strategy. Casinos win in Las Vegas because they have more money and can outlast fluctuations in luck. Similarly, price wars are won by those who can endure the longest due to their financial resources. The owner of a company wields legal power to hire or fire employees. China is a superpower primarily due to its large population, which translates into substantial economic purchasing power. Power stems from force, defined by the quantity of your resources and how effectively you apply them.

Force

The essence of strategy is power, and the essence of power is concentrated force. Power is the effective application of this force. In most cases, when an offense begins, the side with the strongest, most concentrated force prevails. Prussian military strategist Carl von Clausewitz emphasized this: "The greatest principle of war is to always act with concentrated and superior forces against inferior forces."

According to British logician, mathematician, and philosopher Bertrand Russell, force manifests in three forms:

1. **Economic**: Whoever controls the money controls the rewards and punishments. Economic power hinges on the resources you possess, your ability to defend them, your capability to seize possessions from others, and your capacity to receive aid from allies.

2. **Military**: Whoever commands arms and skilled troops can impose or attempt to impose their will on others.

3. **Propaganda**: Whoever influences the minds of the majority shapes public opinion and their perceptions.

Influence can be fleeting, like persuading your spouse to see the movie you want, or enduring, like creating a lasting habit through education, as Nancy Reagan did with her "Just Say No" campaign. Those who control people's knowledge, skills, beliefs, and commitment wield significant power.

The force of your offense or defense is derived from effectively applying your resources to exploit your opponent's vulnerabilities. The impact of your actions depends on the strength of your resources, the speed and concentration with which you deploy them, the extent of your opponent's vulnerabilities, and the robustness of their defense. Ingenuity can compensate for some lack of strength but rarely replaces it entirely.

According to Clausewitz, the application of force is typically not continuous but rather a series of "continuous shocks." Force can be applied directly (hand-to-hand combat) or from a distance (artillery). It can involve individuals (foot soldiers) or machinery (rockets). Force can be mobile (tanks) or stationary (fortifications). It can be captured and used against you, just as you can capture and use your opponent's force. Force is capable of making autonomous decisions, and in planning, you must anticipate all possible decisions your resources might make, including the risk of them turning against you. Force that can make its own decisions can also be persuaded to change direction and loyalties.

The primary adversary of significant force is an even greater force. When attacking intermediate objectives, the challenge is to use the minimum superior force necessary. This strategy conserves reserves and may provoke less resistance. Additionally, applying the least force is quicker and simpler, giving your opponent less time to prepare a defense.

It is crucial to use superior forces in your initial encounter. Winning the first encounter is vital; it confirms the superiority of your resources, undermines your opponent's morale, and boosts your team's confidence for future engagements.

A long-forgotten leader of a religious

commune near Loveland, Colorado, once observed, "Every force will be met with an equal and opposite force." Though this may not have been his original idea, years of observation have proven it true. A corollary to this is another truth: the more you achieve without force, the faster and easier your progress. Avoid using force when unnecessary. The longer you can hold off, the better. When you do use force, apply it at the last possible moment, with maximum strength and concentration in time and space.

Physical Strength

Your resources, which include money, people, territory, allies, equipment, and materials, are crucial in waging battle. While money is vital, having willing people who are spiritually aligned with your quest is even more critical. Building a dedicated and willing force takes time and effort.

Your physical strength starts with your innate personal strength and expands through your influence on others. It relies on your determination, stamina, and available resources to support you and your followers. Accumulating physical resources can be time-consuming. Your strategic location within a territory can also enhance your strength. For example, taking the high ground allows you to shoot down at your opponent, providing more cover for you and less for them. Superior equipment and materials can increase your strength by offering technological and tactical advantages.

Allies

Allies can significantly enhance your arsenal. Sometimes, securing allies requires financial resources. To obtain the money needed to fund your strategy, you have several options:

1. **Earn it**: Acquire money through legal means such as business ventures, investments, or other legitimate commerce. Illegal activities might be considered in some cases, though they come with significant risks and ethical concerns.

2. **Borrow it**: Secure loans or credit from financial institutions, investors, or informal networks. Be mindful of the terms and conditions, as borrowing can lead to debt and economic pressure.

3. **Seize it**: Take money directly from your opponent through force or strategic maneuvers. This approach is risky and often involves significant conflict.

4. **Persuade or Manipulate**: Convince allies to provide financial support. This process can involve negotiation, offering future returns, leveraging relationships, or even deceit.

A robust financial strategy is crucial for sustaining operations and securing the support you need. Allies can provide financial resources and strategic advantages such as additional manpower, expertise, and territorial access. Building solid alliances often involves a combination of economic incentives, shared interests, and mutual benefits.

Remember that an ally chooses to align with you for their reasons, typically driven by self-interest. It is wise to treat allies with a degree of caution. When selecting allies, carefully evaluate the territory they control, the resources they can contribute, and the timing of their potential support. Never rely on allied resources as your primary reserve.

The key to choosing effective allies is ensuring that both parties strongly agree on the underlying principles of your collaboration. Allies are more likely to cooperate if there is a fundamental agreement on core objectives. For example, Islamic countries have often resisted cooperating with Western governments on women's rights because many of their leaders do not fundamentally agree with that principle.

Forming Alliances

Alliances can be invaluable when you need additional resources, complementary skills, or protection from potential attacks. However, alliances should only be formed when critically necessary.

Forming alliances is complex, even when partners share the same culture. When cultural differences are involved, it becomes exceedingly challenging. Cross-cultural negotiations are particularly prone to misunderstandings and misjudgments, both underestimating and overestimating the capabilities and intentions of the other party. Such a lack of awareness increases the potential for failure before negotiations begin.

According to Cynthia Barnum and Natasha Wolniansky in the October 1989 issue of Management Review and Sanfrits Le Poole in Negotiating with Clint Eastwood in Brussels, typical failings when Americans attempt to negotiate in foreign cultures include:

1. **Unfamiliarity with Foreign Surroundings**: Lacking knowledge of the local environment.

2. **Underestimating Distance and Customs**: Failing to account for factors like jet lag, foreign locations, and local customs.

3. **Unfavorable Arrangements**: Negotiating in uncomfortable or unfamiliar settings, such as being seated lower than your counterpart or during prolonged drinking sessions.

4. **Ignoring Personal Circumstances**: Not learning about your counterpart's situation, culture, and language.

5. **Impatience**: Skipping small talk and needing to allow more time for orientation.

6. **Overlooking Areas of Agreement**: Rushing past common ground without thorough exploration.

7. **Premature Concessions**: Making compromises too early in the negotiation.

8. **Time Pressure**: Being constrained by fixed deadlines, such as flight departure times.

9. **Poor Listening**: Not listening attentively.

10. **Misunderstanding**: Listening without fully

comprehending.

11. **Misinterpreting "Yes"**: Especially in Japan, where "yes" may indicate understanding rather than agreement. Or, in cultures where the word "No" is considered rude, like in India.

12. **Rigidity**: Being overly legalistic or competitive.

13. **Misjudging Contracts**: Not realizing that a contract might only signal the beginning of negotiations.

14. **Sequential Issue Negotiation**: Treating each issue as independent of the overall agreement.

15. **Fear of Returning Empty-Handed**: Reluctance to leave without a deal.

16. **Arrogance**: Adopting a "take it or leave it" attitude.

17. **Assuming Fairness**: Failing to recognize that your counterpart does not need to be likable, reasonable, or fair.

18. **Incomplete Analysis**: Overlooking the "fat" in opening offers, missing the broader picture, and not recognizing the significance of certain agreements.

19. **Unprepared for "Salami Tactics"**: Not anticipating incremental changes to the deal after it seems settled, such as last-minute alterations on the way to the airport.

Consider employing a local negotiator who is familiar with the customs and culture. This approach can also give you the advantage of maintaining a level of detachment, allowing you to have the final say in negotiations.

Even with successful negotiations, alliances can be risky. Throughout history, trusted allies have often betrayed their partners for selfish gain. It's crucial to anticipate and plan for such eventualities. One area where relying on allies is hazardous is forming your reserves, as you may find yourself vulnerable when your resources are most needed.

Reserves

The term "reserves" sounds like a military or accounting term, but really the concept of reserves extends beyond those areas. Reserves are resources held back to counter any unforeseen threats. Your rainy-day bank savings are a form of reserves. They are most effective when applied at the critical point, where all force is needed to ensure success. Reserves, when applied, provide relief and a psychological boost.

Balancing Resources for Success

Achieving the right balance of resources is crucial for success. This balance involves a strategic mix of people, supplies, equipment, and sufficient reserves. An illustrative example is from World War II in Europe: General Patton's tanks advanced so quickly that they outpaced their fuel supply line, forcing them to halt until the supplies could catch up.

The condition and readiness of your resources significantly impact their effectiveness. Well-maintained and efficiently functioning resources outperform those that are neglected or broken down. Therefore, your supply line must be meticulously organized, adequately stocked, and always prepared to deliver.

A robust supply line should be:

- Well-rehearsed: Regularly practiced to ensure smooth operations.

- Established: Strongly rooted with clear procedures.

- Stocked: Adequately equipped with necessary supplies.

- Ready, Willing and Able: Motivated and capable of performing under pressure.

- Adaptable: Ready to handle constant change and unexpected challenges.

- Defensive and Dynamic: Able to defend successful strategies while abandoning ineffective ones.

- Risk-accepting: Open to taking calculated risks and managing them effectively.

- Boundary-extending: Willing to go beyond conventional limits when necessary.

- Maintaining this balance and readiness ensures that your operations can continue smoothly, adapt to changes, and overcome challenges effectively.

Concentration

Concentration involves using maximum force in a single action, focused on a single objective at a specific moment. It is a fundamental key to successful strategies. While some individuals and strategies tend to "time-slice"—dividing their time and attention across multiple objectives—this approach often requires additional resources. Each objective faces greater risk, and the overall strategy takes more time.

A crucial lesson from Pearl Harbor is the importance of concentrating your forces during an attack but avoiding concentration at rest or in reserve. Tightly packed resources create easy targets for opponents, and high resource density can indicate the location of leadership.

A superior force becomes weakened when spread too thin, whereas a concentrated inferior force can significantly damage a superior but dispersed force. Similarly, delegation of leadership requires balance; excessive delegation can dilute leadership effectiveness, while too much concentration can stifle initiative.

Human Nature and Resource Psychology

If the great strategists of history had dwelled on human nature's limitations and frailties, many of the past's monumental achievements might never have occurred. Humans are unpredictable, unreliable, prone to mistakes, inconsistent, fragile, and, at times, disappointing. So, how do we ever get anything done? The key lies in collaboration, effective leadership, and doing the best job possible, even if imperfect.

Understanding human nature is not just a helpful tool, but an essential element in designing effective strategies and applying force. Actions depend on the people executing them, who work at different speeds, levels of effectiveness, and degrees of carefulness. This understanding empowers us to tailor our strategies to the unique strengths and weaknesses of our team.

Humans often gravitate towards inertia, immobility, and inactivity. They fall into predictable patterns and routines, resistant to change. People are natural risk-avoiders, preferring others to make the first move. These tendencies can be exploited in your opponent's resources while you work to overcome them within your ranks.

Your state of mind can also be a source of strength. Confidence, preparedness, experience, and skill are all internal strengths. Psychological strength, both yours and that of your followers, begins with your commitment and culminates in your willingness to sacrifice for the cause. Collective psychological strength is bolstered by gains and diminished by losses. Additionally, psychological capital can be gained indirectly, such as by observing your opponent waste resources.

Both physical and psychological strengths are absolute yet relative to your opponent's strengths. Absolute superiority generally leads to success. However, even if your strength is relatively weak, your opponent's strength may be weaker.

Loyalties contribute significantly to psychological strengths. Strong loyalties fortify the organization, magnifying strength whether authority is concentrated or distributed. Your resources' confidence in your leader and their general character is crucial. A distributed force with a strong organization can be as powerful as a concentrated force with a weak organization.

Resource Selection and Application

As we have described, resources encompass any instruments that can be deployed to undermine your opponent's assets in executing their strategy. These resources include the obvious, such as money and people, but also the less apparent, such as time, energy, equipment, supplies, and mental fortitude. While some resources are more durable than others, all eventually wear out and can be damaged. Their usefulness can often be extended or revitalized through rest and repair, depending on the extent of wear and damage. Ultimately, however, all resources become obsolete or irreparable.

Selecting resources should be based on how they will be applied. For instance, Emil Schalk, who authored The Summary of the Art of War for the United States volunteer army in 1862 during the U.S. Civil War, emphasized that resources should be allocated according to their specific abilities and the campaign's needs. He identified three distinct roles for resources based on their unique strengths and the requirements of the strategy:

1. **Infantry** – Versatile and effective at all distances, infantry was the backbone of military power.

   a. **Marksmen** – For distant fighting where skill was critical.

   b. **Bayonets** – For close combat requiring strength and individual courage.

   c. **Reserves** – For decisive fighting involving the strongest and bravest soldiers.

2. **Cavalry** – Best for close fighting, cavalry combined three strengths: the combined strength of human and horse, the speed and shock impact of the horse, and the bravery of the human. Cavalry came in three types:

   a. **Light** – For speed in reconnoitering, outposts, pickets, and foraging.

   b. **Medium** – For moderate force and shock with speed to impact already harassed infantry.

   c. **Heavy** – For maximum force and shock with speed to impact penetrating or retreating infantry.

3. **Artillery** – Best for distant fighting, artillery's strengths were power and precision first and mobility second.

Tools, Expendables, and Rights

In addition to financial and human resources, tools are a significant source of strength. Tools amplify the power of your resources by enhancing their capabilities. The United States, regarded as a leading global power, owes much of its strength to sophisticated tools such as computers, satellites, equipment, weapons, and machinery. Tools are durable, renewable, and repairable. However, like all resources, tools have limits; they require maintenance and can only operate for a certain period before needing replenishment, repair, or replacement.

Maintenance is a continuous process that consumes time, energy, equipment, and resources. In some cases, consumables may be replenished from plundered supplies, but generally, spare parts must be stockpiled.

Some resources are renewable. Human resources, for instance, must be renewed with rest and recovery, which must be planned alongside the transportation needed to move them and the facilities required to house them. Certain cultures and religions emphasize the importance of large families as a long-term strategy for resource renewal.

Expendables are resources consumed and not reusable, such as fuel, food, ammunition, explosives, and supplies. These create waste that must be disposed of safely and sanitarily to maintain the effectiveness of your resources.

Resources can also include intangible assets such as "rights," including patents, trademarks, treaties, alliances, contracts, and other agreements. These can attract additional resources, protections, and opportunities.

Information as a Source of Strength

Information and the communication channels through which it travels are critical resources and sources of strength. Accurate, unfiltered data about your objectives and your opponent's resources and goals is essential. Analyzing data to turn it into useful information requires experience and knowledge about analysis, your resources, and your opponents. Experienced analysts are crucial for this process.

Effective communication hinges on listening and understanding. If you can understand your opponent better than they understand you, you have a significant advantage. But if you can understand your opponent better than they understand themselves, you hold the key to prevailing in most contests.

Information provides a competitive advantage. But it's not just about having the information; it's about recognizing and acting on a goal before your competition does. This preemptive action is what allows you to reach the goal first and secure your advantage. For example, in the early 1970s, United Airlines and American Airlines began offering their computer reservation systems to travel agencies. United automated many agencies quickly, but American targeted the most profitable agencies and sold them their system before United could. By identifying the true goal first, American secured a larger market share than their size would have typically allowed. Later, when United bought an accounting system to enhance their services, American bought the company that developed the software, limiting United's upgrade access. In both cases, American's superior analysis and preemptive actions led to success, despite United being first to market.

A critical category of information is the assessment of your opponent. Consider the following in your assessment:

1. Number of resources, their freshness, and morale

2. Vulnerabilities

3. Flow of material and people

4. Preparedness—the degree to which they are ready and practiced

5. Awareness—their degree of mental alertness

6. Willingness—assessment of their commitment

7. Their objectives inside and outside your territory

8. Their degree of penetration into your territory

9. Endurance—ability to go far, fast, and to last

10. Experience—depth of their hands-on learning

Strategic Use of Information

Attacking an opponent who is fresh, strong-willed, and prepared is far more challenging than targeting one who is exhausted, disorganized, and disheartened. Experience is a valuable form of information. If you lack experience, you will need more information. Veterans bring a wealth of information to your team, including better self-control in conflict or danger, superior judgment in decision-making, and enhanced skills that make them more effective.

Information reduces risk, substantiates plans, and provides the foundation for choosing one approach over another. Gathering information can range from formal espionage to modern market research.

Political campaigns never stray far from their research. Politicians use focus groups, surveys, polls, and audience meters before, during, and after every debate. They carefully research public opinion to craft messages that resonate, often practicing short, impactful quotes ("sound bites") to convey their positions quickly. Given the public's short attention spans, messages must be clear and concise. However, if elected, these carefully crafted messages may reflect something other than the candidate's intentions; they are strategic moves to win office.

Information flow is bidirectional. How information is disseminated is as crucial as how it is received. While the message must serve a purpose, it does not always need to be entirely accurate. For example, consider the Willie Horton case. Horton, described by some Republicans as a 'black murderer' and 'rapist,' was released from a Massachusetts prison under an early release program. This portrayal exploited racial prejudices and was strategically used to generate public and political controversy. In his article "Why It Was So Sour" in the November 14, 1988 issue of Time magazine, Walter Shapiro explained how George Bush Sr.'s staff used this information in the Republican campaign against Michael Dukakis. Dukakis, the governor at the time, was labeled a "liberal" and, by extension, responsible for the release of dangerous criminals.

Notably, the fact that Dukakis's predecessor created the early release program was conveniently omitted from the TV spot, and Dukakis's reasons for defending the program were never discussed in the press. This example illustrates how presenting information based on facts but twisted, incomplete, or distorted can be damaging.

Poignant messages with subliminal content work well in political contexts. However, such approaches can backfire. In the 1992 U.S. presidential race, Bill Clinton admitted to trying marijuana in college to convey honesty but faced ridicule when he claimed he "didn't inhale."

One critical success factor in strategic thinking is applying the right resources, such as people and information, in the right amounts to achieve one's objectives. This approach requires careful planning.

5. Planning for the Anticipated

In addition to having a clear idea of your motive and your objectives as we described above, before you begin to lay out the steps that are obvious and necessary to reach your objectives, you must first have these requirements for planning your strategy.

1. Some idea of how much time you have available for planning

2. A handle on the key assumptions upon which your plan will be built

3. A certain degree of genius

4. The right information gathered about your opponent

Allocating Time for Planning

How much time should you spend on planning? The time available for planning depends on several factors, with the primary consideration being whether you are planning an attack or defending against one. If you are defending against an attack that has already begun, it is too late for detailed planning. However, if you are preparing for an offensive or defending against a future attack, more time should be allocated to planning, especially when significant resources are at risk, and the degree of risk is high.

If the consequences of failure are minor, you can afford to spend less time planning. In your planning process, be mindful of avoiding two common human tendencies:

1. To have that last bit of information

2. To rush and act without information

Your plans must be as thorough as time permits, but they must not be so exhaustive that they consume more time than you can afford. It is crucial to strike a balance between thoroughness and efficiency. Continually evaluate your plan from the perspective: "Can I proceed and feel reasonably confident with my plan?"

Key Assumptions for Building Your Plan

Unless you have definitive information, you will need to make several critical assumptions during your planning. For instance, assume that the territory where you will be operating is unfamiliar. Anticipate encountering maximum resistance at the most crucial moment—when you must deploy your peak level of resources, self-confidence, experience, and skills. Additionally, consider your actions if your opponent discovers your plans in advance.

Testing the sensitivity of critical assumptions is a vital step in your planning process. You will undoubtedly make many other critical assumptions. For these, evaluate the potential outcomes if the assumption proves incorrect, in either direction. Never rely heavily on an assumption without verifying and double-checking its validity.

Genius

It's a misconception that you need to be highly intelligent to excel at this fundamental form of planning. What's truly important is being quick, thorough, and thoughtful. Clausewitz's assertion that superior intellect is crucial for a successful strategy has led to this misunderstanding. He believed that someone with superior intellect does not go against the natural order of the laws of probability, and that superior intellect is the result of energetic genius. However, his views do not imply that you must be smarter than your opponent or have more information.

Clausewitz suggests that genius in strategy planning can arise from hard work and thorough analysis. Genius is when you mentally synthesize all the facts into a coherent organization, similar to when you sleep on a problem and wake up with the solution.

Clausewitz also meant that genius is a product of unwavering determination to succeed. It's not about genetic inspiration, flashes of brilliance, or the pursuit of novelty; it comes from hard work, preparation, and most importantly, mental flexibility. This adaptability is what allows you to adjust your strategy when circumstances change.

Guided by Clausewitz's insights, your planning should consider the following:

1. **Conflicting Motives and Aims**: Determine whether your motive or aim and that of your opponent are fundamentally at odds, ensuring no compromise is possible.

2. **Resource Strength and Positioning**: Assess how well your resources' strength and location align with those of your opponent, identifying any advantages or disadvantages.

3. **Resource Readiness**: Evaluate your resources' readiness, considering their fitness, training, attitudes, abilities, and skills.

4. **Political Sympathies and Impact**: It's important to consider the political sympathies within other territories regarding your endeavor. Understanding the potential effects your aim may have on them will make you more aware of the broader implications.

Information in Planning

Information is a fundamental requirement for planning. However, even with ample information, planning the application of resources remains more of an art than a science. Each decision requires the right amount of information—sufficient but not overwhelming. You need information about the intermediate objectives, obstacles along the way, and the past effectiveness of your resources in similar situations. Similar insights into your opponent's circumstances would also be valuable.

Your resources must be tailored to the unique needs of your specific intermediate objective, environment, and opponent. Training and drilling your resources to ensure they are always ready is crucial. Manage the flow of resources, maintain your tools and resources in top condition, and create maximum strength by concentrating and focusing your resources on the narrowest aspect of your intermediate objective while maintaining control throughout the process. Selecting the right initial point of attack is often essential to the success of your strategy.

Actions can take on a life of their own, so carefully consider the consequences of your planned actions. Avoid over-planning or agonizing over the sufficiency of a plan. Judge the thoroughness of a plan based on the amount of resources at risk—the greater the risk, the more thorough the plan should be. As Emil Schalk, who summarized the art of war during the Civil War, stated, "The advantage falls to whoever acts first with the greatest energy and coolness and to whoever has studied and planned the best beforehand."

Additionally, remember Clausewitz's insight: "Wars are the result of one side, if not both, miscalculating their chances for success—for if they both could do so with certainty, there would be no war."

What to know

Some of the information you ideally want to know about your opponent includes the following:

1. What is, or likely wll be, your opponent's objective(s)?
2. What is the size of your opponent's forces?
3. What forces will you need to overcome those of your opponent?
4. Where are your opponent's forces

concentrated, and how are they distributed?

5. How might your opponent act or react from their present position?

6. Where are your opponent's communications lines, lines of retreat, and supply lines?

7. Which of your opponent's areas are most accessible to attack? To defend?

8. How is your opponent using the natural cover of the surrounding terrain?

9. What approach will provide the speediest attack?

10. How will the first victory lead to the second?

11. How will time and distance be managed?

12. How soon will your opponent's subsequent reinforcements arrive?

13. What can you do to neutralize your opponent's allies?

You can gather information through various methods. Public records often hold valuable data; some information might be directly observable from your current position. Additionally, employing a spy or detective can help acquire more elusive details.

One approach is to train a spy from within your own ranks to infiltrate the enemy's territory or organization, though it carries significant risk. If discovered, the spy could face severe consequences, and deeply embedded organizations, like Al-Qaeda, may be particularly difficult to penetrate. This underscores the need for caution and thorough planning.

Alternatively, you might recruit a spy who is already on the inside by using bribery, blackmail, or threats to coerce one of your enemy's assets into working for you. It is crucial to ensure the loyalty of such a spy, as there is always the risk that they could be turned into a double agent, ultimately working against you.

Protecting your information from enemy spies is of utmost importance. This can be achieved through encryption, physical security measures, or by compartmentalizing information so that no single individual has access to the entire message—much like how pirates divided treasure maps.

If you suspect an infiltrator within your ranks, one effective strategy is to plant different pieces of enticing but false information with various individuals. By monitoring which false data is leaked, you can identify the spy.

What Needs to Be Planned

For your plan to be complete, you must consider the following aspects:

- **Application of Force**: Determine where, when, and how you will deploy your resources against your opponent, including the number of forces and their specific roles.

- **Sequence of Events**: Outline the step-by-step progression of your actions to ensure a coherent and coordinated effort.

- **Control Mechanisms**: Establish methods

to control your operations and swiftly address unforeseen changes or challenges.

- **Logistics**: Plan for procuring, transporting, and distributing necessary supplies and equipment.

- **Training**: Ensure all participants are adequately trained and prepared for their respective roles.

- **Communications**: Develop a robust communication strategy to facilitate clear and efficient information exchange among all team members.

- **Timing of Launch**: Decide on the precise moment to initiate your plan, considering both internal readiness and external conditions.

- **Risk Management**: Identify potential risks and devise effective contingency plans to mitigate them.

- **Intelligence and Reconnaissance**: Gather and analyze information about your opponent to inform strategic decisions and anticipate their actions.

- **Evaluation and Adjustment**: Implement an ongoing assessment and feedback system to adapt and refine your plan.

Force

According to Clausewitz, you should plan to apply your force with the "utmost concentration, the utmost speed, and without interruption." Speed is a critical component, but impact is equally important. During World War II, General Patton demonstrated this principle by using specially designed tanks to move rapidly and decisively against German forces. Mechanized units can outpace infantry, airplanes can outpace ground vehicles, and telephone communications can outpace airplanes.

Fidel Castro's maneuver during the Bay of Pigs invasion is a historical example of appropriate speed and impact. He positioned tanks swiftly by having them pre-loaded on flatbed trucks, allowing them to reach strategic positions faster than they could.

In planning the application of force, it's essential to consider the rate at which your resources will be consumed. This includes personnel, fuel, food, and equipment, all of which will need to be replenished and repaired. However, equally crucial is financial planning—determining the required funds, managing cash flow, and assessing the associated risks. This aspect of military strategy is often overlooked but is a key factor in the success of any operation.

You should plan the use of your resources according to their inherent strengths and characteristics. Each type of resource has its most effective application, and recognizing these distinctions will empower you to optimize your strategic outcomes. This understanding is a powerful tool in the hands of any military strategist.

For example, during the Civil War, according to Emil Schalk, infantry was divided into distinct categories based on their roles and skills:

1. **Marksmen**: These soldiers relied on precision and skill, engaging in combat from a distance.

2. **Bayonets**: This group required individual courage and physical strength as they fought in close-quarters combat.

3. **Reserves**: The strongest and bravest soldiers were held back as reserves, ready to be deployed for decisive engagements.

Cavalry was classified into the following categories:

- **Light Cavalry**: Utilized for their speed in reconnaissance, establishing outposts, picketing, and foraging.

- **Medium Cavalry**: Combined moderate force and shock with speed to effectively impact already harassed infantry.

- **Heavy Cavalry**: Applied maximum force and shock with speed to decisively engage penetrating or retreating infantry.

Consider that within any organization, whether a family or a government, the use of force is regulated by a set of rules or laws enforced from the top down. One effective strategy for gaining control over money, people, and territory is to secure the resources and then build loyalty through a combination of favoritism and discipline.

Sequence of Events

When people think of strategy, they naturally focus on the planning aspect. Planning a strategy involves determining a series of steps to achieve intermediate objectives, ultimately leading to the final goal. Each step in the plan must include decisions about what actions to take, what resources are needed, how to position those resources, and what support will be required along the way. While each step should be logical, it doesn't have to be predictable. The plan must also outline how progress will be tracked and reported and how success will be measured. Emphasis should be placed on achieving the intermediate objectives.

At a minimum, any strategic plan should answer the following questions for each intermediate objective:

1. **Sequence Placement**: Where does the objective fit in the overall sequence of events?

2. **Value Contribution**: What value will the objective contribute to reaching the final objective?

3. **Resource Allocation**: What types and amounts of resources will be applied, expected, and maximum?

4. **Control Mechanisms**: How will control be maintained throughout the process?

5. **Resource Application**: How will resources be applied to achieve the objective?

6. **Timing of Resource Application**: When will the resources be applied?

7. **Reserve Resources**: What types and amounts of resources will be held in reserve, and where will they be located?

8. **Expected Results**: What results are expected from achieving this objective?

9. **Contingency Plans**: What contingency actions can be taken for unexpected setbacks or outcomes?

10. **Communication Plan**: How will progress and any issues be communicated to all relevant stakeholders?

11. **Performance Metrics**: What specific metrics will measure the objective's success?

12. **Support Requirements**: What additional support or coordination with other teams or departments will be needed?

13. **Risk Assessment**: What potential risks are associated with this objective, and how will they be mitigated?

14. **Feedback Mechanisms**: How will feedback be gathered and used to adjust the strategy if necessary?

When devising a strategic plan, it is crucial to anticipate your opponent's movements. If your opponent is on the move, you must aim your strategy ahead of your opponent's projected position when you launch.

Contingency Planning

When planning your strategy, you must meticulously plan how to purchase, accumulate, store, transport, and distribute all your supporting supplies. This planning involves designing lifelines for supplies, fuel, food, materials, equipment, replacements (both people and parts), and all other necessities for your resources once they are deployed. As your resources move, your supply lines must keep pace.

External factors, such as random events and emotional impacts, can disrupt your sequence of events. Unforeseen incidents or irrational acts by your opponent can cause events to occur out of sequence or prevent key steps from being accomplished. Control can be unexpectedly lost due to random events; for example, a rainstorm can slow your progress. Similarly, emotions can have unpredictable consequences. In an American football game, a taunt might provoke a punch, resulting in a penalty that could affect the game's outcome and potentially the entire season.

When planning the sequence of events for your strategy, it's crucial to control your movement from each step to the next, while also anticipating future steps and providing contingencies. It's equally important to plan for every possible consequence of your strategic actions and those of your opponents. Remember, in critical situations, rely on tried and true resources—avoid untested resources where success depends on their flawless performance.

Your plan will differ between an internal strategy (such as a revolution) and an external strategy (such as a war). In an internal struggle, distinguishing friends from enemies can be challenging. Allies are more complicated to trust, and the distance your opponent needs to cover to reach you is shorter. In an external struggle, the enemy is more easily identifiable, allies are less prone to deception, and the distances are farther, providing more time and potentially more warning.

Equally important is planning for reassembly after the attack—designating a meeting point to account for all detached groups and organize them into new units. Additionally, it's crucial to anticipate dealing with the aftermath and how life will proceed after your success or failure. Comprehensive planning, considering all possible outcomes, is critical.

Logistics

Your plan must encompass logistics, detailing the time and methods to obtain, assemble, and secure the necessary resources, including personnel, materials, fuel, food, and more. Additionally, the plan must allocate time for training and practice and implement measures to prevent spying and information leaks. It would be best to plan for your security as the leader.

When planning your attack, consider all potential directions from which a counterattack might come.

Emil Schalk described the logistics during the U.S. Civil War as follows:

1. **Preparation**: Prepare your materials and position them strategically for the opening campaign.

2. **Reconnaissance**: Deploy reconnaissance teams to confirm the enemy's location.

3. **Security**: Establish guards to secure all directions.

4. **Communication**: Ensure that lines of communication are established and reliable.

5. **Mobilization**: Initiate the movement of your army.

Strategic Communication Planning

Effective communication planning involves determining what information will be sent, to whom, where, when, how often, why, and how it will be transmitted. The old parlor game demonstrates how communication can deteriorate as it passes from person to person. Designing a shallow organizational structure can mitigate this problem by shortening communication lines, allowing information to travel more quickly and accurately. A shallow organization can also move swiftly and cover more ground in less time. In such a structure, leaders function more independently and are more likely to use their initiative when events do not unfold as planned.

The flow of information to and from your center of operations is crucial. The quality and security of communications depend on factors such as distance, frequency, message length, susceptibility to interception, and the importance of the information.

While it may seem that having dual sources of information on the same subject would improve accuracy, it's important to be aware of the potential impact of human factors. Competition between sources, cooperation before reporting, or attempts to disprove each other can all affect the quality of the information. Discrepancies between reports can cause delays until resolved. It's tempting to trust common information from both sources, but it's important to remember that both could be wrong. Therefore, it's crucial to view communications with suspicion and seek additional ways to confirm critical facts.

When you suspect that your communication lines are vulnerable to interception, it's crucial to take precautions to maintain the integrity of your communications. Consider sending false information or encoding your messages. However, it's important to remember that both the sender and recipient must understand which messages are valid and which are not. Every received message could be tampered with, contain lies, errors, or exaggerations, be incomplete, or be based on faulty assumptions. Therefore, it's essential to verify critical facts and not take any communication at face value if the matter is serious.

Communication serves several critical purposes, including:

1. **Reporting**: Provide updates on the condition or status of resources, ensuring that all parties are informed about current capabilities and needs.

2. **Fact Consolidation**: Group and consolidate facts to create a clear and cohesive understanding of the situation.

3. **Connecting Knowns with Unknowns**: Link known information with unknown variables to identify gaps and facilitate problem-solving.

4. **Associating Concerns with Facts**: Relate current concerns to facts to contextualize issues and guide decision-making.

5. **Reevaluating Connections**: Disconnect previously connected facts when new information invalidates old assumptions, ensuring that strategies remain accurate and relevant.

6. **Coordination**: Facilitate coordination between different teams and departments to ensure a unified approach to achieving objectives.

7. **Decision Support**: Provide decision-makers with accurate and timely information to support strategic and tactical decisions.

8. **Feedback**: Collect and relay feedback from various sources to continually refine and improve operations and strategies.

9. **Alerting**: Warn about potential environmental threats or changes that require immediate attention.

10. **Motivation and Morale**: Communicate goals, successes, and encouragement to maintain high morale and motivation within the team.

Disrupting and Securing Communication

Disrupting communication can be achieved by pressuring or punishing the communicator, adding noise to distort or drown out the message, cutting off the communication medium, or removing the person at either end of the communication. Preventing communication can be essential to executing a strategy.

Your plan must be comprehensive, specifying not only the disruption tactics but also where you will position your resources, including their targets, instructions, the locations of command centers, and communication methods during your attack. General Patton believed in keeping his leaders close to the front lines to maintain short communication lines, obtain firsthand intelligence on the battle's progress, and provide a brave role model for the troops.

Choosing your form of communication and channels should prioritize speed and security. Whether the communication is written on paper, carried by horseback, or sent electronically via secure email, anticipate potential interception or disruption. Historically, various means such as carrier pigeons, dogs, submarines, satellites, and dolphins have been used to carry messages, each with varying degrees of success. Ensure you have multiple means of communication and at least one secret backup channel.

Importance of Training

Endurance, skill, pride, and high morale are cultivated through rigorous training. While training is a partial substitute for experience, it is often the best preparation available before engaging with your opponent. Training equips your resources with the necessary knowledge and skills. Experience, on the other hand, imparts intuition, wisdom, and discernment to know when to be bold and when to be cautious. Effective training, which is crucial in developing teamwork and independent operation, ensures that you are prepared and adaptable for any situation.

Training, as emphasized by Clausewitz, can achieve the following:

1. **Decision-Making Foundation**: Equip your resources with the skills to make informed decisions in unfamiliar circumstances, saving time when decisions are straightforward.

2. **Insight and Confidence**: Extend mature insight to lower ranks, which becomes increasingly important with more hierarchical levels. Training guards against eccentric actions and mistakes by instilling methods and routines crucial when time is limited. It also fosters a sense of preparedness among your resources, giving them the assurance that

they are well-equipped to face any challenge.

3. **Effective Leadership**: Enable brisk, precise, and reliable leadership—reducing natural friction, eliminating hesitation, and smoothing the way forward. Training provides a basis for consistency and discipline, ensuring a cohesive and effective team.

Training can inadvertently make your resources rigid, unimaginative, and prone to panic when faced with unfamiliar situations. It is crucial to include handling the unexpected as part of their training to prevent these drawbacks. Proper training can mitigate fear during execution, improving the quality of information received. Fear tends to distort and exaggerate information, leading to inaccuracies in reports. Additionally, training conditions your resources, building their endurance, which ultimately saves time during execution.

Launch Timing

Launch timing is crucial when initiating your first offensive action and throughout your entire campaign. Consider the classic gangster movies where the gang huddles around a diagram of the bank, discussing their assignments for the upcoming heist. The last thing they do is synchronize their watches to ensure everyone is in their assigned place at the right time. Your planning should be equally precise.

Timing is typically aimed not at achieving surprise but at coordinating resources to arrive at the right time, focusing their strength to create power, anticipating when reserves will be needed, and having those reserves in place at the critical moment.

Understanding your opponent's tendencies provides valuable insights into what to expect from them. Are they early adopters? Might they overreact to a feint? Are they methodical and thus a bit slow to make decisions? Is your opponent a short-term or long-term strategist?

By accurately timing your actions and understanding your opponent's behavior, you can ensure your resources are optimally deployed and your strategy remains effective throughout the campaign.

Timing is crucial when deciding when to launch your strategy. Clausewitz believed that the optimal time to begin is indicated by one of the following conditions:

1. **Vulnerable Opponent**: When your opponent is vulnerable, meaning they are ill-fitted and unprepared to defend.

2. **Plan Strength**: It's crucial to be in a position of strength and readiness before launching your strategy. This means being well-prepared, having the strength to win, and the conditions being well-suited for the execution of your plan.

3. **Plan Alignment**: When your resources match up well against the distribution of your opponent's resources, and your opponent is weak, indecisive, or making errors.

4. **Urgency**: When the need for action is

imperative.

It's the strategist's responsibility to recognize these signals, as it ensures that you launch your strategy at the most advantageous moment, thereby maximizing your chances of success.

Plan to pounce when the timing is right. The time required for success will be shortened if you have more resources than your opponent, move more quickly, and strike with more concentration. Speed is essential; it prevents defenses from being set up, forces your opponent to make hasty decisions, lessens your overall cost, and increases your opponent's cost. The rate of deployment is also critical; it can range from an all-out attack, as in "Charge!" to small doses, such as snipers.

All of your "strength extenders," such as tools, fitness, morale, attitude, and information, multiply the strength of your resources, will accelerate your progress, and will help to ensure your success.

Another aspect of timing is time of day. People function differently at different times of day; some people function better in the morning while others do better late at night. Some fade just before mealtimes, while others fade on Fridays, after a hard week at the office. How people respond to proposals can vary by time of day.

Timing is also related to control. If you keep the right pressure on your opponent, your opponent will be responding to your timing. If your opponent is controlling the flow of conflict, then you are subject to your opponent's timing.

Once you have planned what needs to be planned, as described above, your strategy should be nearly ready but not quite.

Your plan must anticipate the entire timeframe—from the opening salvo through the consequences of your success or failure. America's initial military incursion in Iraq was impressive but the planning for reconstruction after the end of the organized hostilities appears to have been severely under-planned.

Even if you have anticipated all the possible risks and opportunities and planned for them, you must not allow your plan to deny the impact of superior numbers. If your opponent has twice as many resources, you will need to recognize your diminished likelihood of success.

And, of course, you must recognize that no matter how well you plan, something unanticipated is likely to happen.

Maintaining Control in Strategic Planning

A significant element of strategic planning is designing a plan for always maintaining control of the situation. If control is lost, your offense can quickly turn into defense. Therefore, you must create a plan that ensures control and anticipates the unexpected as you proceed step by step.

In the ancient game of Go, players alternate placing stones on a nineteen-by-nineteen grid. The

objective is to get five stones in a row. If you have four in a row with open squares at both ends, you can win on your next turn—unless your opponent is similarly positioned. The game thus becomes a battle for getting three in a row with open ends.
Throughout the game, control shifts, and you can feel the momentum change whenever one player gains or loses control.

Control is the least understood aspect of strategy, which is critical to success. The degree to which you control a situation determines your chance of success. At one extreme, having no control means you are a victim of circumstances. At the other extreme, maintaining total control throughout the strategy's execution ensures a higher likelihood of success. You can achieve this by recognizing key "points of control" and steadily progressing from one to the next, maintaining balance. Progressing steadily means controlling each intermediate objective while moving at a pace that allows maximum speed to the next objective.

In the middle ground, you may sometimes have control and lose it at others. It is possible to regain control and succeed, but losing control even once can lead to failure. American football is a good example: when momentum shifts, control is lost. A failed first down results in a punt, giving the opposing team a chance to drive and shift control.

Control typically has two states: either you have it, or you do not. If you lose control, your opponent gains it. In rare circumstances, control can be taken from both opponents by an act of God (such as an accident or natural disaster) or by intervention from a more powerful third party. While you do not plan to lose control, you must plan how to proceed if you do.

Recognizing key "points of control" and managing your progress from point to point is a skill that can make you great. For example, politicians carefully control the development of public opinion. They avoid haste, build consensus, and progress steadily by aligning objectives and maintaining control.

Another example of "points of control" is navigating physical pathways. Controlling key pathways can be an intermediate objective that leads to control of your final objective, such as cutting off your opponent's resources.

Understanding your opponent's movements is crucial. Study the network through which you and your opponent will travel. Test your opponent's defenses. For instance, in Vietnam, the North Vietnamese made several small attacks before their main assault to learn how the Special Forces A-Team would react.

Some strategies target individuals rather than large groups or organizations. Control remains a crucial element. A well-planned strategy keeps control throughout its execution, regardless of intermediate outcomes. One way to control your opponent's behavior is to set a mode of behavior for yourself and conditions for your opponent that leaves them with limited choices. Communicate this behavior persuasively to your opponent while limiting their choices by providing conditional alternatives. For example, saying, "I am going to get my gun, and if you are still here when I come back, I'm going to shoot you," limits your opponent's choices to:

Leaving

Staying and risking being shot

Leaving and returning with some form of defense

By defining these limits, you maximize your advantage and control the range of possible outcomes.

6. Planning for the Unanticipated

You can anticipate many potential events through careful planning. By brainstorming all the possible things that could go wrong, you can identify situations that warrant preparation and those that are so unlikely that you may choose to address them only if they occur.

Managing Chance in Strategic Planning

A chance event is, by definition, completely unpredictable. Chance is the one factor you cannot plan for, cannot rely on, and should not depend on. You can plan for every possible outcome to minimize the impact of chance, but chance events will still occur.

While chance is statistically predictable in the long run, it is not predictable in the short term. For example, when flipping a coin, you know that tails will come up half the time over many flips, but you cannot predict the outcome of the next flip. Your best strategy is to identify as many chance factors as possible and have contingency plans to address their potential impact. Once engagement begins, remaining flexible and creative is the best protection against chance events. Well-defined principles, policies, and guidelines will assist your field commanders in adapting to deviations from the plan.

Imperfect information about your opponent's strategy is a significant variable. The likelihood of your information being wrong is high. However, wrong information only becomes a chance event if you act on it. Here, logical deduction and intuition play a crucial role in reducing the element of chance, providing you with a sense of reassurance and confidence in your decision-making.

Being in the right place at the right time is usually not a matter of luck but of planning. To increase the likelihood of your resources being in the right place at the right time, you must be aware of where the right place will be. Control your actions to position your resources in the vicinity of the right place at the right time, and be prepared to seize opportunities as they arise. This proactive approach will make you feel ready and in control. You can determine the right place by studying your opponent's motives, behavior trends, recent activities, resource locations, and general human nature.

As an example of planning to be at the right place at the right time, we once knew a beautiful woman in her early twenties who had recently divorced a famous sports figure. She confidently said she intended to marry a wealthy man when asked about her future. After some laughter, she explained that she had researched where wealthy single men vacationed. She arranged a six-month leave of absence to go to those places and find a suitable match. Her plan worked—she married a wealthy music producer with homes in London and New York City.

Understanding and Mitigating Risks

Risks are threats to your resources that can be predicted to some degree. While you cannot eliminate all risks, you can mitigate their impact. One example of risk is the information you collect on your opponent. What you believe you know about your opponent may be fraught with misrepresentations, incomplete analyses, and outright falsehoods.

Another significant risk is the possibility of inaccuracies in assessing your opponent's strength. Clausewitz noted that while it might seem advantageous for a small group to defeat a larger one, what typically occurs is an underestimation of the opponent's strength. As a result, a smaller force may encounter more resistance than it can handle.

By recognizing and preparing for these risks, you can better navigate the uncertainties inherent in strategic planning and execution.

Sun Tzu made the following observations about risk:

1. Every Battle is Won Before it is Fought: Victory is achieved through superior strength and meticulous planning.

2. Know Your Enemy and Know Yourself: Understanding your enemy and your capabilities is crucial for success.

3. Fight for a Moral or Noble Cause: Instill belief in the cause and in command discipline to motivate your forces.

4. Minimize Engagement: The objective is not to

devastate the enemy but to win with as little engagement as possible, breaking the enemy's resistance without fighting (though General Patton might disagree).

5. All Warfare is Based on Deception: Deception is a fundamental strategy in warfare.

6. Avoid Prolonged Operations: A clever operation is rarely prolonged.

7. Maintain Unity: Keep your group united, govern in the people's interests, and maintain national unity.

Risk can be effectively managed through concealment and secrecy. Avoid predictability by using feints and bluffs, and gather as much information as possible about your opponent.

The Role of Surprise in Strategic Planning

Incorporating surprises into your plan can magnify your force, intimidate your opponent, and reduce their morale. However, Clausewitz noted that surprise is most effective when time and space are limited in scale. Small surprises have the best chance of success but are less likely to lead to significant victories.

The primary benefit of a surprise attack is that it limits your opponent's ability to strike back effectively. A well-executed surprise should still succeed even if it becomes known beforehand.

It is important to remember that detaching a

portion of your force to execute a surprise can weaken your main force. If the success of your plan hinges on the element of surprise, it may indicate that your plan is too risky.

Secrecy

Concealing your plans and keeping their location a secret are essential to your strategy. Concealment involves hiding information or resources, while secrecy entails keeping quiet about their location and protecting your concealed information from discovery. Secrecy also involves hiding information and movement. There are five degrees of secrecy:

1. **Secret:** Currently being protected from becoming known.

2. **Believed Secret**: Not likely to be known, but a remote possibility exists.

3. **May Become Known**: Some chance of becoming known.

4. **Will Be Known**: Likely to become known when it becomes important.

5. Known: Formerly concealed information has become known.

It is crucial to keep your plans secret and disguise your purpose as much as possible. Sometimes, you may even need to conceal certain parts of your plans from your subordinates. However, if your subordinates become aware of this concealment, it can convey a lack of trust and create morale problems and suspicions. This situation can lead to your resources becoming distracted, losing momentum, and potentially shifting their loyalty.

Failings

According to the article "1990 Access Issue" in the first quarter 1990 edition of Dowline, typical failings in planning include:

1. **Stopping Short of Optimum**: This occurs when you fail to consider all possible alternatives. People often stop when they find a suitable solution, missing out on potentially better options. Always search for multiple alternatives before evaluating each one.

2. **Ignoring Personal Values and Motives**: It is essential to consider the personal values and motives of everyone involved, including your people, your allies, and your opponents.

3. **Overlooking Illogical Actions**: While you may consider specific actions illogical, your opponent might find them logical. Additionally, incorporating an element of illogic into your own plan can confuse your opponent.

Being this thorough can lead to the planning excess we cautioned against. You can minimize the impact of excessive thoroughness by planning quickly and efficiently.

Final Check

Once you have your best plan laid out, review it thoroughly by considering the following:

1. **Assume It Is Discovered**: How will you respond if your plan is discovered? Develop contingency plans to adapt quickly and minimize the impact if your strategy is compromised.

2. **Evaluate Random Scenarios**: Create and simulate random scenarios to assess how they would be handled. This trial run helps identify potential weaknesses and improves your readiness for unforeseen events.

3. **Plan for Information Gaps**: Consider what would happen if crucial information is unavailable at a critical time. For example, what will you do if, in the crucial moment, you do not know if a critical event (e.g., XYZ) has occurred? Establish protocols for decision-making under uncertainty.

4. **Anticipate Opponent's Actions**: Estimate the impact if your opponent targets all of your weaknesses. Develop strategies to protect your vulnerabilities and respond effectively to an aggressive counterattack.

5. **Rank and Address Risks**: Rank the areas of risk from highest to lowest and ensure that the highest risks have been anticipated. Allocate resources and create contingency plans to mitigate these high-risk areas.

6. **Streamline Execution**: Look for ways to

accelerate the execution of your plan, removing any unnecessary steps that could slow you down. Efficiency can be the key to maintaining momentum and swiftly achieving your objectives.

7. **Consider Collateral Damage**: Assess the impact of collateral damage to innocent bystanders and critical infrastructures (e.g., bridges) and the handling of prisoners. Develop ethical guidelines and strategies to minimize and manage collateral damage responsibly.

8. **Plan for Casualties**: Determine how you will manage your casualties in the short and long term. This management includes medical care, evacuation procedures, and maintaining morale among your forces.

9. **Secure Leadership**: Ensure leaders are positioned where they can be secured while clearly viewing the engagement. Decide whether you need to be distant, establish an alibi, or close to the action for firsthand information. Plan communication channels to keep leaders informed and capable of making timely decisions.

10. **Ensure Flexibility**: Incorporate flexibility into your plan to adapt to changing circumstances. Train your team to respond to new information and unexpected developments while maintaining cohesion.

11. **Review Resource Allocation**: Double-check that resources are allocated efficiently

and can be quickly reallocated if needed. Ensure supply lines are secure and capable of supporting rapid changes in your strategy.

12. **Rehearse Critical Actions**: Conduct rehearsals for critical parts of your plan. Practice essential maneuvers and decision points to ensure everyone understands their roles and can execute under pressure.

13. **Assess Long-Term Implications**: Consider the long-term implications of your actions. Plan for the aftermath of your strategy, including political, social, and economic consequences. Ensure that your plan supports sustainable success.

14. **Establish Clear Objectives**: Revisit your objectives to ensure they are clear, measurable, and achievable. Ensure every part of your plan aligns with these objectives and contributes to your overall goal.

7. Execution--Offense

Strategy is complex, and at its core lies offense. Offense initiates a strategy; there is only the status quo without it. According to your plan, you must execute quickly, accurately, crisply, and decisively. Aim to win the war in the first battle, or if that's not possible, win the first and then the war in the second. Offense involves exerting your control or will over that of your opponent. This action can mean taking or destroying their resources or driving them from the control of an intermediate or final objective.

The decision to start is always the hardest decision to make. When decisions are difficult, indecision can creep in. Indecision is not much different from consciously deciding not to act. Likewise, choosing not to act can sometimes be indecision in disguise. Once the decision is made, offense becomes a state of tension—maximum tautness.

Once your execution begins, do not let fears creep in. Research by Mathew Bothner, a Graduate School of Business faculty member at the University of Chicago, revealed that NASCAR drivers are most likely to crash when they fear the driver behind them is about to pass them. Do not overreact to threats or let fear distract you from your strategy. Stick to your plan unless you learn something dramatically different from your planning assumptions. Do not let a setback stop your initiative. Find a way around it. If necessary, "burn your ships" as Cortez did, ensuring that your resources have no means to return and must keep moving forward.

According to Clausewitz, among the qualities of intellect, determination, and presence of mind, determination is the most important. Determination drives action when all other motives falter.

Execution consists of the following stages:

1. **Massing**: Gathering your resources. This stage involves consolidating all necessary personnel, equipment, and supplies to ensure readiness for action. Effective massing requires careful planning and coordination to maximize the strength and availability of your assets.

2. **Deployment**: Positioning your resources. Once gathered, your resources must be strategically positioned to maximize their effectiveness. Deployment involves moving resources to specific locations, ensuring they are ready to engage at the right time and place.

3. **Engagement**: Engagement is the fight in which resources are consumed. Direct action takes place in this phase, and resources are actively utilized. Engagement requires precise execution, adaptability, and resilience, as it is the stage where the outcome of your strategy is most directly contested.

4. **Regrouping**: This stage involves collecting your remaining resources. After the engagement, it is essential to regroup and reassess. This step consists of consolidating

remaining resources, assessing losses, and preparing for subsequent actions. Regrouping is critical for maintaining momentum and adapting to new circumstances that arise from the engagement.

Each of these stages is vital for the successful execution of a strategy. Proper planning and coordination at every stage ensure that your resources are utilized efficiently and effectively, increasing the likelihood of achieving your objectives.

The Spectrum of Offensive Strategies

The range of offensive strategies is vast, from friendly persuasion to violent destruction. You can gently convince an opponent to accept your will, or you may launch an offense that utterly destroys the stabilizing forces in their lives, even against their most robust resistance. Many familiar forms of offense are between these extremes, from simple games to the intricate dynamics of love and courtship. Offenses vary in the degree to which they apply pressure to remove your opponent's independence—the freedom to choose their path. Many forces compete to strip individuals of their independence, including those who attempt to control their money (e.g., stockbrokers, Wall Street) and their thoughts (e.g., newspapers, TV, religions).

One offensive strategy is negotiation. However, negotiation is not always an option, as it depends on the open-mindedness and willingness of all involved parties. With two parties, there are four possible combinations of agreement: agree-agree, disagree-agree, agree-disagree, and disagree-disagree—of which only the first has both parties willing to agree. Often, an offense is launched when negotiation fails. Negotiation is preferable to violence when fundamental human rights are not at risk, as in the American Revolutionary War, where negotiation with Britain was no longer a viable alternative.

At the same time, being too pliable can signal weakness, lack of motivation to fight, or absence of firm principles. In such cases, your opponent may sense your weakness and impose their will on you. According to the Executive Strategies newsletter, the keys to successful negotiation are the following:

1. **Preparation**: Thoroughly understand your position and your opponent's position.

2. **Clarity**: Clearly articulate your goals and desired outcomes.

3. **Flexibility**: Be willing to make concessions while holding firm on critical points.

4. **Listening**: Actively listen to understand the interests and concerns of the other party.

5. **Patience**: Allow time for discussions to unfold without rushing the process.

6. **Respect**: Maintain a respectful demeanor to build trust and rapport.

7. **Creativity**: Seek innovative solutions that satisfy the interests of both parties.

Leadership

Leadership, as described in *The Art of the Leader* by Dr. William A. Cohen, centers around personal strengths essential for effective leadership. These strengths include:

1. **Strength of mind** -- encompasses self-control, self-confidence, courage under stress, and intuitive wisdom gained through experience.

2. **Firmness** -- characterized by unwavering principles. These principles are not rigid but are founded on consistent thought, reflection, and adherence to proven principles. A firm leader is also compassionate, fair, reliable, and trustworthy, striking a balance between strength and empathy.

3. **Effective decision-making** -- combines thoroughness, agility, and the ability to make sound judgments swiftly. A leader capable of decisive action in crises can prevent myriad pitfalls, unnecessary risks, and subsequent corrections.

4. **Understanding** -- includes a deep knowledge of adversaries, current issues, and subordinates' strengths, virtues, and shortcomings. This acute ability to assess both people and operational needs is a crucial strength in effective leadership.

5. **Creativity** is the capacity to devise innovative solutions to persistent problems or novel approaches to emerging challenges.

6. **Endurance** is the capability to sustain prolonged efforts, nurtured through thorough preparation.

7. **Resilience**, the ability to rebound from significant setbacks, also fostered by adequate preparation.

A leader must often make difficult decisions, such as removing key personnel when dissent arises, performance falters, or disruptions disrupt team cohesion.

Understanding your opponent is crucial. Their willpower, ingenuity, and decisiveness determine their endurance, creativity, and effectiveness as a leader. However, it's vital to recognize that your opponent may not share your mindset, values, ideals, or biases. They may consider options that you wouldn't and vice versa.

While essential in life, certain personal traits like honesty, integrity, ethics, and honor may hold little sway in dealings with an opponent. If these values matter to your opponent, leveraging them could work to your advantage, especially if you don't possess those traits.

Leadership is a learnable skill. While specific innate characteristics like personality and physical stature (such as height) can aid in leadership, they aren't prerequisites for effective leadership. Standing at five feet six inches, Napoleon exemplified this, as did Mother Teresa, who led despite her advanced age and frailty. Ron Howard, known for his role as Opie on The Andy Griffith Show, also demonstrates leadership qualities in his career as a film director.

One of the most remarkable leaders I've encountered was Colonel Curtis "Colt" Terry, a five-foot-eight Special Forces officer who honed his leadership skills alongside his comrades in Korea and Vietnam during his twenties. He led through example, cunning, and unwavering loyalty to his team. The depth of his leadership was evidenced by the enduring loyalty of those who served under him throughout their lives. For further insights into his leadership, refer to my biography of him.

Dr. Cohen outlines the following principles for aspiring leaders:

1. **Individual Impact**: Emphasizing that one person can indeed make a significant difference.

2. **Collaborative Success**: Stressing the importance of leveraging teamwork and motivating others to perform at their best to achieve shared objectives.

3. **Informal Leadership**: Highlighting that leadership can manifest without a formal leadership title or position.

4. **Consistency in Leadership**: Asserting that effective leadership transcends situational factors and external conditions.

Stepping into a leadership role, a challenging yet achievable endeavor according to Dr. Cohen involves the following actions:

1. **Initiative**: Willingness to assume leadership and take charge of situations.

2. **Risk Acceptance**: Embracing and managing risks effectively.

3. **Responsibility**: Accepting and responsibly managing any delegated authority and power.

4. **Innovation**: Fostering creativity and exploring new ideas.

5. **Expectation Setting**: Establishing high standards and goals for oneself and others.

6. **Positive Attitude**: Maintaining an optimistic and proactive "can do" mindset.

7. **Visibility**: Being actively present and visible, leading from the forefront.

As a leader, you wield the ability to distribute rewards, enforce sanctions, impart crucial information, inspire through encouragement and recognition, and offer essential support and assistance to your team.

Dr. Cohen outlines seven crucial steps for assuming leadership in crises:

1. **Set Clear Objectives**: Define your goals clearly to guide your team effectively.

2. **Effective Communication**: Communicate calmly, clearly, and with impact.

3. **Bold Action, Not Recklessness**: Act decisively while considering consequences.

4. **Timely Decision-Making**: Make and communicate decisions promptly, without waiting for perfect information, to avoid delays, changing circumstances, missed opportunities, competitors' advantage, and loss of team morale.

5. **Assert Control**: Take proactive measures to gain and maintain control, anticipating developments and surpassing self-imposed limitations.

6. **Lead by Example**: Demonstrate willingness to perform tasks you expect of others.

7. **Personnel Management**: Effectively manage your team by replacing underperformers discreetly.

General George S. Patton famously emphasized the importance of direct communication in leadership, stating, "If you can't get them on the phone, then you shouldn't be giving the orders, and if you can, then it is much better to do it in person. This is not at all burdensome because few people call a general except in cases of great emergency, and then they like to get him at once."

In organizational dynamics, power is typically bestowed from higher authorities, while influence among people is earned through demonstrated competence and trustworthiness. A successful leader focuses on achieving goals and outcomes rather than fixating on the methods used. Over time, effective leadership—even if perceived as harsh or severe—can be justified by its results. Conversely, leaders who fail to achieve their objectives often face criticism regardless of their leadership style, whether perceived as harsh or benevolent.

Dr. Cohen suggests that leaders can effectively attract and retain followers by implementing the following practices:

1. **Awareness**: Staying informed about current events and internal dynamics.

2. **Supportiveness**: Offering assistance to those in need.

3. **Collaboration**: Seeking help from knowledgeable individuals when necessary.

4. **Problem-Solving**: Identifying and addressing the root causes of issues.

5. **Opportunity Recognition**: Seeking out and leveraging new opportunities.

6. **Respectful Feedback**: Publicly praising team members while reserving criticism for private settings.

7. **Vision Sharing**: Clearly communicating your goals and aspirations to inspire others.

8. **Facilitating Understanding**: Helping team members understand each other and the broader mission.

9. **Encouraging Healthy Competition**: Using competition to motivate and improve team performance.

People act based on their own motivations, not yours. As a leader, striking a balance between giving clear, firm directions and nurturing enthusiasm for shared goals is not just crucial, but also inspiring. Your actions and how you lead are significantly more influential than how your subordinates follow, and this should motivate you to foster a shared enthusiasm for the goals you set.

In essence, effective leadership rests on eight core principles: possessing a clear vision, asserting leadership, setting high standards, fostering innovation, embracing calculated risks, maintaining a positive outlook, being a visible figure, and genuinely caring for each team member. Adhering to these principles is fundamental to becoming a successful leader. And, other characteristics should be considered.

Leader Smartness
Leaders can exhibit effectiveness through a diversity of intellectual strengths, and their adversaries may possess a type of intelligence that differs from theirs. Recognizing and adapting to the variety of ways intelligence can manifest is crucial for effective leadership.

There are several distinct ways in which individuals can demonstrate intelligence, each valuable in different contexts:

1. **Street Smart**: This type of intelligence is derived from real-world experiences rather than formal education. Street-smart individuals often have a practical

understanding that can give them an edge in everyday situations, especially against those who lack the same hands-on experience.

2. **Book Smart**: Knowledge acquired through formal education and studying the experiences documented by others. This type of intelligence provides a theoretical foundation and depth of knowledge that can predict and understand an opponent's moves if their strategies are derived from similar studies.

3. **Creative Genius**: This form of intelligence is marked by inventiveness and the ability to synthesize known information into new, original ideas. Creative geniuses are often unpredictable, capable of devising unique solutions that may catch more conventional thinkers off guard.

4. **Incisive Smart**: Characterized by an ability to identify the heart of an issue quickly and efficiently, cutting through irrelevant details. Such individuals remain focused on the core goals, making them formidable opponents in scenarios that require direct and decisive action.

5. **Analytical Smart**: Reflective and systematic, these individuals think deeply about information before acting. Their systematic approach often results in well-thought-out plans that, while potentially rigid, are thoroughly devised to cover anticipated scenarios. They may struggle, however, when

situations change unpredictably.

6. **Generalist Smart**: This type combines elements of all the types above. Generalists have a broad knowledge base and are adaptable, blending analytical depth with creative breadth. They can devise complex strategies incorporating innovative and unpredictable elements, making them incredibly versatile and challenging opponents.

Each type of intelligence has its strengths and potential weaknesses. Effective leaders, by embodying or harnessing these various forms, can enhance their leadership style and strategy.

A compelling demonstration of intelligence was observed in a study involving an octopus in Washington State. In this study, a researcher placed a shrimp inside an open jar in a tank with an octopus, a known delicacy for the creature. The octopus explored the jar with its tentacles, discovered the opening, and promptly retrieved and ate the shrimp. In a subsequent trial, the researcher secured the jar with a lid. Undeterred, the octopus felt around, located the lid, and ingeniously figured out how to unscrew it to access the shrimp inside.

The experiment didn't stop there. While the researcher prepared for another round, the octopus, anticipating the next move, climbed into the jar itself, correctly predicting the location of the next shrimp. This incident not only illustrates the octopus's ability to solve problems but also highlights its capacity for anticipatory thinking, demonstrating a level of intelligence that we can find both surprising and inspiring.

Leader Bravery

Dr. William A. Cohen's research offers insights into the motivations that drive individuals to step forward in crises.

Here's a detailed exploration based on his findings:

1. **Dependence of Others**: Individuals often take action because they know others are relying on them, recognizing their role as crucial within the group.

2. **Fear of Judgment**: Some are motivated by the fear of being seen as cowards or facing other negative social judgments if they fail to act.

3. **Loyalty to Leadership**: A strong belief in and respect for their leaders can drive individuals to act to meet expectations and earn approval.

4. **Perceived Safety in Action**: For some, moving forward may seem safer or more strategic than staying behind, driven by calculating risks.

5. **Training and Reflex**: Extensive training can condition individuals to respond automatically to crisis, with actions deeply ingrained by repeated practice.

6. **Desire for Recognition**: The anticipation of rewards or recognition can motivate individuals, fulfilling their need to feel valued and important.

7. **Personal Growth**: Some see crises as opportunities for personal development and growth, stepping forward to stretch their capabilities.

8. **Proof of Competence**: Individuals may be driven to prove their skills and worth to others, viewing the crisis as a test of their abilities.

9. **Thrill-Seeking**: The inherent excitement and adrenaline rush associated with dangerous situations can appeal to some.

10. **Altruism and Patriotism**: A deep sense of duty or selfless desire to contribute to the greater good can inspire action, often rooted in strong personal or cultural values.

These motivations are complex and can vary widely among individuals, often overlapping in real-world situations.

Leader Execution

Your presence and supervision at the onset of strategic operations are critical as a leader. Being on-site allows you to maintain order and focus during the initial phase of any hostile action. It's essential to keep the momentum going once you initiate action.

Halting your advance gives your opponent an opportunity to recover, regroup, and counterattack, potentially shifting the momentum against you.

If circumstances force a pause, it's essential to resume action slowly and cautiously to regain momentum safely. Restarting too quickly can lead to hasty decisions, potentially leading your forces into unanticipated dangers, such as traps set by the opponent. The key is to maintain a speed of action that outpaces your opponent, which helps you reach your objectives more swiftly and limits your opponent's ability to strategize and respond effectively. By controlling the pace, you deny your adversary the time to think through their actions and set up defenses.

Actions in strategic operations can be understood as a cyclical process involving five key phases:

1. **Resources Deployment**: This initial phase involves positioning your resources strategically to optimize their effectiveness.

2. **Active Engagement**: Here, you engage directly with the opponent's forces. Typically, superior resources will overpower weaker ones, but exceptional strategies, innovation, or psychological tactics can sometimes lead to an underdog's success.

3. **Withdrawal**: This stage occurs after resources are significantly depleted or objectives have been met. Withdrawal may signify a retreat due to losses or a strategic repositioning after achieving goals.

4. **Reassessment and Regrouping**: After engagement, it's time to learn. Reassess the situation to understand the progress made and any adjustments needed. This phase is crucial for learning from the engagement and preparing for future actions, making you an active part of the strategic operations.

5. **Return to Planning**: With insights gained from the reassessment, it's time to learn and prepare for the following action phase. This might involve selecting a new objective or refining strategies, thus restarting the cycle. This phase ensures you are always prepared and ready for the next step.

This repeating cycle ensures that strategy is dynamic, adapting to changing conditions and learning from ongoing engagements. The notion of strategy being executed in "pulses" reflects the active and reflective periods inherent in strategic operations.

The term "pace of progress" was coined to describe the speed at which an organization moves toward its goals, such as achieving profitable sales growth. This concept highlights the importance of assessing and managing how quickly a company can reach its strategic targets.

Several key factors influence the "pace of progress" within an organization:

1. **Resource Allocation**: Effective use of resources on new initiatives is crucial. Each investment should have a distinct purpose that propels the organization toward its strategic goals.

2. **Streamlined Structure**: Minimizing bureaucracy and empowering lower-level decision-making can significantly accelerate processes. In such an environment, directives move swiftly from top to bottom, and the organization is not overly controlled by cautious financial oversight.

3. **Risk Tolerance**: Maintaining a willingness to accept calculated risks can drive innovation and progress, allowing the organization to seize opportunities that more risk-averse competitors might pass up.

4. **Customer Focus**: Focusing closely on the customer's needs and benefits ensures that all actions are aligned with delivering value, enhancing customer satisfaction and loyalty, which in turn drives business success.

The agility and effectiveness of small businesses are often attributed to several key factors:

1. **Speed and Flexibility**: Smaller units generally operate more swiftly and adapt more easily than larger organizations, benefiting from less bureaucracy and more direct communication paths. However, they may lack the extensive resources that larger entities possess.

2. **Managing Impatience**: Recognizing and managing impatience is crucial. It can lead to starting projects prematurely, depending too much on quick fixes, and panicking when plans falter. However, when strategically exploited in competitors and carefully managed within one's team, impatience can be a powerful tool.

3. **Controlled Momentum**: Maintaining balanced momentum is crucial in executing strategies. It's vital to move quickly yet remain vigilant against potential pitfalls like overextending or falling into competitors' traps.

4. **Monitoring Structure**: Continuous monitoring of how resources are organized, their strength, interaction between units, and alignment with strategic goals is essential.

5. **Monitoring Performance**: It is critical to evaluate the effectiveness of these resources in achieving specific objectives. This involves comparing current outcomes to the expected results outlined in plans and, if available, past

performance.

6. **Cost-Benefit Analysis**: Always assess the progress made against the costs incurred. This analysis helps determine whether the strategy yields sufficient returns relative to the resources and efforts expended.

7. **Ordered Retreat**: If retreat becomes necessary, it should be executed orderly to ensure security and minimize losses, maintaining balance and protection for all involved.

By following these principles, you can enhance your organization's operational efficiency and strategic effectiveness, particularly in high-pressure or rapidly changing environments.

Leader Plam Maintenance

The concept of entropy in physics—stating that order in the universe naturally declines—can serve as a compelling metaphor for organizational management. Like physical structures and biological systems, organizational order and efficiency inevitably deteriorate over time. Therefore, it is crucial to actively manage and frequently reassess the organizational structure and strategy to maintain effectiveness.

Here are some improved approaches and insights:

1. **Continuous Reassessment**: Just as entropy increases over time, leading to decay, organizations must continuously fight this decline by revisiting and revising their strategies. This reevaluation should be done periodically, especially during the execution phase of any plan, to adapt to changing circumstances and correct course as necessary.

2. **Managing Stress and Decisions**: Operating in a state of offense often introduces high tension and stress, exacerbating the potential for error in decision-making. It's vital to have a robust, well-defined plan that can guide your team through stressful periods and ensure decisions are not made in haste.

3. **Balancing Offense and Defense**: While your research indicates a prevalence of offensive strategies over defensive ones, the simplicity and effectiveness of defensive strategies should not be underestimated. Although generally fewer and more straightforward, an effective defense can be crucial and often requires more strategic planning and complexity than offensive maneuvers.

4. **Strategic Flexibility**: The inherent unpredictability and complexity of offensive and defensive strategies demand flexibility and the ability to pivot quickly in response to unforeseen challenges or setbacks. This agility

is essential for maintaining a competitive edge and achieving strategic goals.

By acknowledging the natural decline in organizational order due to entropy, leaders can better prepare to counteract this trend through vigilant planning, strategic flexibility, and periodic reassessment. This proactive approach, when consistently applied, is not just a strategy, it's a mindset that is essential for sustaining long-term effectiveness and success in any organization, giving you the confidence to manage organizational effectiveness.

Leader-Recap

Taking all the above factors into consideration, your execution typically starts with offenses. A list of specific offenses appear in the chapter that follows.

From my research, I found that there are more offensive strategies than defensive ones. Fewer defenses exist because they tend to be more effective and are generally simpler than offensive strategies. However, this simplicity does not diminish the inherent complexity of many offensive maneuvers.

Center of Gravity—Base of Operations

Your strategy originates from your base of operations, which could be either mobile or stationary, such as a political headquarters. Defending this central hub is essential, as it serves as the main gathering point for personnel and the nexus from which intelligence is analyzed and strategic decisions are disseminated. Moreover, you might establish subsidiary bases for specific tactical initiatives to support targeted operations effectively. This approach ensures that your primary and auxiliary bases are well-coordinated and safeguarded, maintaining the integrity and efficacy of your operations.

When selecting a location for your base of operations, it is crucial to consider several strategic factors to ensure optimal functionality and security. Here are the initial points expanded with additional considerations:

1. **Distance**:
   - **Resource Deployment**: Assess the logistics of moving supplies, personnel, and information. The base should be ideally positioned to minimize the time and cost of transportation.
   - **Communication Efficiency**: Consider the impact of distance on communication speed and reliability with deployed resources.
2. **Security**:
   - **Risk of Attack**: Evaluate the likelihood of the base being targeted by adversaries.

- **Vulnerability of Supply Lines**: Ensure supply routes are secure from disruption and sabotage.
- **Defensibility**: Choose a location that can be easily defended against potential attacks, considering natural and constructed fortifications.
- **Surveillance Capabilities**: The ability to monitor surrounding areas effectively can provide advanced warning of potential threats.

3. **Mobility**:
- **Relocation Readiness**: The base should allow for rapid evacuation or movement if the security situation deteriorates.
- **Secondary Locations**: Plan for alternative sites that can serve as temporary bases if the primary location becomes untenable.

4. **Accessibility**:
- **Access for Allies**: Ensure that allies can easily reach the base if their support is needed.
- **Geographical Constraints**: Consider natural barriers like rivers, mountains, and deserts that might hinder access to the base.

5. **Sustainability**:
- **Resource Self-Sufficiency**: Evaluate the local resources for sustainability—like water, food, and energy supplies.
- **Infrastructure**: Existing structures, road networks, and other infrastructures can influence the choice of location due to ease of adaptation and use.

6. Legal and Political Factors:
- **Jurisdictional Considerations**: Understand the legal implications of establishing a base in a particular location, including compliance with local laws and regulations.
- **Political Stability**: Assess the political environment of the area to avoid locations in volatile regions which could complicate operations.

Considering these factors thoroughly will help establish a secure, effective, and resilient base of operations against various threats and challenges.

People are inherently territorial, often defending what they construct or possess with vigor. Typically, they establish a primary center of operations, which effectively becomes their hub of power or center of gravity. In strategizing an attack on their territory, identifying this central hub is crucial—it's where their resources, command, and control are most concentrated, and from where strategic directives are issued.

This center of gravity is not just the heart of their operational strength; it also represents the point from which the most significant threat to you could be launched. It is the focal point for their offensive actions and a retreat for recuperation. To effectively counteract your opponent, focus on understanding their actions—these often provide clearer insight into the location and nature of their central strength than their words might reveal. Targeting this hub can disrupt their operations and diminish their ability to mount or sustain an offensive, thus shifting the balance of power in your favor.

According to Clausewitz's warfare principles, effectively targeting your opponent's center of gravity involves a strategic approach that maximizes your forces' impact while minimizing their exposure and vulnerability. Here's how to apply these tactics:

1. **Identify Central Sources**: Analyze and trace your opponent's strength to as few sources as possible, ideally pinpointing a single central source. This will help you focus your efforts effectively.

2. **Assess Attack Feasibility**: Consider the practicality of deploying your maximum force against their principal source of strength. This consideration involves evaluating your resources and the opponent's defenses.

3. **Minimize Distance**: Calculate the shortest route to the center of gravity. This knowledge will minimize exposure and maximize the element of surprise and operational efficiency.

4. **Stay Secure: maintain Concentrated Forces**: Avoid any strategic plans that require dividing your forces. Splitting strength typically dilutes impact and increases vulnerability. By keeping forces concentrated, you can ensure your security.

5. **Swift, Direct Action**: Advance towards the center of gravity along the quickest path at the highest speed. To ensure your force remains concentrated, choose an approach the

opponent least expects to enhance the strike's effectiveness.

Adhering to these guidelines can strategically disrupt your opponent's capabilities while conserving your resources and maintaining a robust offensive posture.

By "unexpected approach," we refer to either an unconventional direction or method of attack. T.E. Lawrence's tactical move during the capture of Al'Aqabah, as described in his autobiography Seven Pillars of Wisdom and popularized in the film Lawrence of Arabia, exemplifies this strategy. Lawrence led his forces across a formidable desert—choosing a route that the defending Turks had presumed impassable due to its harsh conditions—to attack from the landward side where the city's defenses were weakest, achieving surprise and trapping the enemy against the sea.

As you move toward the center of gravity, it's crucial to set and achieve interim objectives that are strategically valuable, attainable, and defendable while also safeguarding your vital assets. Clausewitz emphasizes maintaining constant contact with the enemy to prevent idle time during transit and to keep pressure on their operations.

Attacking directly at the opponent's center of gravity is most effective; if they retreat, they lose the advantage of their stronghold. Should they regroup there, their options become limited, especially if you can cut off their supply lines and maintain continuous pressure. Clausewitz advises that the most significant damage should be inflicted during the opponent's retreat, destroying their resources comprehensively.

In a business context, a company's center of gravity could be a dominant market share controlled through extensive retail presence or critical personnel. Strategic moves like hiring away key employees from a competitor bolster your company's capabilities and weaken the competitor's operational effectiveness.

Individually, your center of gravity might be your professional credibility, support network, or authoritative position. Attacking any of these can significantly undermine your influence and shift power to your opponent. Recognizing and reinforcing these areas can prevent critical losses and maintain your competitive edge.

Transportation

Transportation is crucial in distributing resources, supplies, and reinforcements as a vital link between production and deployment. The efficiency of transportation systems hinges on speed and capacity—faster modes with higher carrying capacities ensure rapid deployment. Choices range from slow, traditional methods like donkey carts to rapid, modern transport like jet planes, each with specific requirements and operational challenges.

Critical aspects of transportation include:

1. **Vehicle and Medium Requirements**: Different modes of transportation, such as

donkeys or planes, have distinct needs, such
as food and water or fuel and landing strips.
The travel medium, whether roads or air, also
presents unique challenges and requirements.

2. **Impact of Terrain**: The nature of the
terrain significantly affects transit time.
Rougher routes increase travel time and the
risk of delays or attacks, necessitating greater
focus on route maintenance and security.

3. **Matching Load to Vehicle**: To optimize
resource efficiency, the cargo's size, weight,
and importance must be aligned with the
vehicle's capacity and the travel medium.

4. **Fuel Considerations**: Transporting fuel
itself requires fuel, highlighting the need for
careful planning to avoid shortages, especially
during delays.

5. **Unforeseen Delays**: Mechanical failures,
adverse weather, and other disruptions can
delay transportation and affect the condition
of the cargo. Perishables can spoil, and
sensitive items can be damaged, underscoring
the importance of contingency planning.

6. **Strategic Importance of Punctuality**:
Timely delivery is crucial not just for the
condition of the cargo but also for the
recipients' readiness and effectiveness.
Investing in excess capacity can serve as
insurance against potential disruptions.

7. **Vulnerability of Transportation
Systems**: In strategic contexts, such as

military operations, attacking an opponent's transportation capabilities—explicitly targeting the mediums (like roads or airways) and then the vehicles—can cripple their ability to sustain operations.

Effective transportation management ensures a steady, reliable supply chain, vital for maintaining operational momentum and achieving strategic objectives. As seen during emergencies, like severe weather forecasts, supply disruptions can lead to immediate and visible public reactions, such as panic buying. Thus, maintaining a robust and flexible transportation system is essential for both routine operations and in response to crises.

Source of Strength

Understanding the origins of your opponent's strength is crucial, as it may differ from their center of gravity. This source could range from external allies to various funding mechanisms, such as illegal operations or secure financial reserves like Swiss bank accounts. Identifying and severing these sources can significantly weaken your opponent's ability to sustain operations.

To effectively target these sources, it's crucial to prioritize disrupting the leadership first. They are typically key to maintaining the operational integrity and strategic direction of the organization. For instance, law enforcement agencies often employ tactics like plea bargains with lower-level offenders to dismantle criminal networks, aiming to work their way up to the leaders or kingpins. This strategy has proven to be effective, instilling confidence in its potential to dismantle the opponent's strength.

By targeting the leadership and their critical sources of support, you can significantly undermine and potentially neutralize your opponent's operations. This approach is a common strategy in law enforcement and military contexts, aimed at disrupting the command structure and logistical support to diminish an opponent's operational capacity and strategic effectiveness. For instance, law enforcement tactics often include negotiating plea deals with minor offenders in drug networks, encouraging them to provide information that can lead to the capture of higher-level figures, such as kingpins. This method weakens the network by removing its leaders and disrupts its operational capabilities by severing critical sources of strength.

Crucial Time

When planning to attack the center of gravity, it's crucial to time the assault for when your opponent is most vulnerable—whether due to distraction, weakness, or lack of preparedness. Choosing both the timing and the location of your attack maximizes your strength and preparedness, enhancing your chances of catching your opponent off-guard.

Even if your opponent anticipates and prepares for the attack, striking with full force at the right moment is essential. Always engage with your maximum capability, ensuring that reserves are ready to support the main effort. This approach prevents the risk of underestimating the opponent's resilience, which can lead to failure.

The importance of full-force engagement is underscored by historical military lessons, such as the early strategic errors made by the CIA in Vietnam. They split Special Forces A-Teams into smaller units, believing that smaller teams could manage the same objectives effectively. This mistake often led to increased vulnerability and mission failure. In combat, as in any strategic operation, every role—from leadership to support like medics or communications—is critical, and reducing these capacities can compromise the entire operation. Always approach with the full strength of your team to maintain tactical superiority and mitigate risks.

Launching

Launching an offensive requires careful consideration of several key factors:

1. **Your Readiness:**
   - **Self-Assessment:** Before launching an offensive, evaluate whether your motivation is strong enough to sustain the effort through potential consequences. A sense of moral duty, a desire for revenge, or power might drive this motivation.
   - **Preparation:** Consider if you are prepared both strategically and resource-wise. Always aim to be in a state of readiness, as this minimizes the disadvantage when an unexpected confrontation occurs.

- **Assessment of Opponent's Vulnerability:** Understand your opponent's current state and potential resilience. This evaluation should include evaluating their strengths, weaknesses, and readiness to respond to your actions.

2. **Public Opinion:**
    - **Influence of Public Sentiment:** Public opinion can significantly impact the success of your offensive. As noted by philosopher Bertrand Russell, public sentiment often shapes societal norms and laws. An offensive against these norms risks public backlash unless it can sway public opinion.
    - **Strategic Communication:** It is crucial to manage how your supporters and the broader public perceive your offensive. Gaining support can provide legitimacy, practical assistance, and reduced resistance.
    3. **Impact on Local Populations:** If the local population supports your cause, they can offer invaluable support. Conversely, if they oppose your actions, they may hinder your efforts.

Each of these factors plays a vital role in the planning and executing of an offensive strategy. Understanding and navigating them can determine your endeavors' ultimate success or failure.

Advance Warning - Yours

For the preservation of your resources, you want to make sure that you will have advance warning of any impending attack. You can do this by putting out "security." For a fixed location, security would consist of an outpost, sentry, or lookout in each direction from which an opponent could attack. Their warning and any resistance that they can offer will give you the time to prepare before opposing forces arrive. If your resources are on the move, you should send out advance guard resources ahead of your main force, to both sides, and depending on how fast you are moving, to your rear. The advance guard also serves as your rear guard in the event your forces must retreat.

Advance Warning - Theirs

A notable trend in recent years is the practice of giving innocent civilians warning of an impending attack. This strategy serves multiple purposes, and its use depends on your objectives.

1. **Humanitarian Considerations**: Providing warning demonstrates a commitment to minimizing civilian casualties and upholding humanitarian principles. It can help to evacuate non-combatants from the area, thereby reducing the loss of innocent lives and potential collateral damage.

2. **Psychological Impact**: Warnings can psychologically impact both the civilian population and the opposing forces. It may reduce panic and chaos for civilians by giving

them time to seek safety. For the opposing forces, it can create uncertainty and disrupt their defensive preparations, knowing that an attack is imminent.

3. **Public Relations and Moral High Ground**: Issuing warnings can enhance your standing in the eyes of the international community and your populace. It shows that you are considering ethical considerations, which can be a powerful tool in the information war. Maintaining the moral high ground can also help in post-conflict scenarios, facilitating smoother reconstruction and reconciliation processes.

4. **Tactical Disadvantages**: On the other hand, providing a warning can also have tactical disadvantages. It gives the opponent time to prepare or evacuate critical assets, potentially reducing the attack's effectiveness. Opponents might also use the warning period to fortify their positions, making the offensive more challenging and costly.

5. **Operational Security**: Warnings can compromise operational security. The element of surprise is a critical factor in many offensive operations, and providing a warning can negate this advantage. The decision to issue a warning must balance the ethical implications with the potential loss of strategic surprise.

6. **Context-Dependent Decision**: The decision to warn civilians should be

contextual. In some scenarios, such as urban warfare, where civilian casualties could be extremely high, the ethical imperative might outweigh the tactical disadvantages. In other cases, where strategic objectives necessitate maintaining the element of surprise, the decision might lean towards not issuing a warning.

In conclusion, the decision to give civilians advance warning of an impending attack is complex and requires careful consideration of ethical, tactical, and strategic factors. The mission's overarching objectives, potential humanitarian impact, and operational context should guide this critical decision.

Lines of Attack

Emil Schalk, a noted 19th-century military theorist, outlined several tactical approaches for conducting military attacks, each suited to different strategic needs and battlefield conditions:

1. **Direct Assault**: This approach involves a focused, head-on attack where forces move forward in a tight formation, similar to driving through a narrow pass. This tactic concentrates power in a single direction, maximizing force but also risk.

2. **Skirmish Line**: This formation spreads the troops across a broad front, extending from left to right. This tactic, used historically by

the British in the Revolutionary War and by Allied Forces during the Normandy beach landings in World War II, allows for a broad engagement front, covering more ground and engaging multiple enemy positions simultaneously.

3. **Flanking Maneuvers with Parallel Columns**: In this tactic, forces are divided into multiple columns that move in parallel, attacking the enemy from several directions simultaneously, particularly from the sides. This tactic can overwhelm an opponent's ability to respond effectively across multiple fronts.

4. **Multiple Columns with Numerous Engagement Points**: Similar to flanking maneuvers but involving even more columns, this strategy creates a dynamic front of attack, resembling waves or ribbons of forces converging on the enemy. This method disperses the enemy's focus and can penetrate defenses at multiple points.

The effectiveness of each formation depends on terrain, enemy deployment, and campaign objectives. Adequate protection with well-coordinated front, rear, and flank guards is crucial to prevent encirclement or ambush.

Choosing the right attack formation is crucial and should align with an overarching strategy that considers the strengths and vulnerabilities of one's own forces and those of the opponent.

8. Execution—Defense

Overview of Defense Strategy

As conceptualized by Carl von Clausewitz, defense fundamentally involves "awaiting the blow." Its primary objective is to conserve resources while leveraging the strategic advantage of geographical positioning. Defense serves to protect and can act as a launching pad for future offensive actions. The primary goal of defense is to repel the enemy, not necessarily to defeat them. The goal is to hinder the enemy and prevent a decisive outcome.

While defenses share some similarities with offenses, such as reliance on strength, timing, focus, attitude, and environmental considerations, many are distinct. Defenses differ from offenses in that their primary purpose is to delay the enemy or protect against the loss of territory rather than to seize it.

Key Elements of Defense

- Strategic Positioning: Choose locations that provide natural obstacles to the enemy, making it costly for them to attack. Such positions include high grounds or narrow passes which force the enemy into a disadvantageous approach. Don't back yourself into a corner or put your back to obstacles (eg., water, cliffs, etc.)

- Resources: Plan to have the most resources you can in case supply lines are cut.

- Resource Conservation: Situate your resources strategically to create a buffer between the enemy and your critical assets, reducing the likelihood of significant losses in an assault.

Time

Time as a Defender's Advantage: Time generally benefits the defender, allowing for preparation and fortification. However, prolonged periods may also give the enemy time to gather more resources.

Detailed Defensive Strategies
1. **Leveraging Terrain**: The terrain should be used to naturally disadvantage the enemy, and complement your defensive strength, especially when your forces are outnumbered. A strong location can make up for inferiority in strength. When defense is necessary, pick a location that narrows the enemy's size down to that of yours, with obstacles to slow them down.

2. **Dynamic Resource Management**: Manage your resources to adapt to shifting battlefield conditions. Keep significant forces in reserve to reinforce positions dynamically or counterattack when opportunities arise.

3. **Deflection**: Position your resources to deflect the force of your opponent's power. As in judo, in defense you use your opponent's force against them. By using their momentum with balance and timing, often you can tip the direction of their force to steer them to a more vulnerable position.

4. **Managing Divided Forces**: Even when outmanned, if your opponent divides, then you divide. Keep more than half your strength in reserve. If your force is divided, move the reserves back and forth to alternately reinforce each part. In this way, you should be able to regain any ground given up before the reinforcements arrive.

5. **Force Exhaustion Management**: Understand the three causes of force exhaustion—fatigue of personnel or equipment, depletion of supplies, and loss of personnel. Plan for adequate recovery and resupply to sustain defensive operations.

Exhaustion and Depth in Defense

1. **Immediate Exhaustion**: Address exhaustion swiftly as it can cripple the defense's effectiveness. This involves rotating forces and managing logistics to ensure continuous capability.

2. **Strategic Depth**: Use depth in defense to manage confusion and disarray, especially in prolonged engagements. This depth should be planned to allow for effective fallback

positions and secondary lines of defense. More depth is needed where more confusion is likely to occur.

Transitioning from Defense to Offense

1. **Utilizing Defensive Gains**: Shift to offense when the enemy's forces are depleted or when a strategic vulnerability is identified. When the front line retreats, the reserves should advance for morale, to reduce disarray, and to inflict maximum damage.

2. **Decision Points for Offensive Transition**: Schalk said, "The invading army becomes smaller the more it advances, while . . . the defending army generally gets stronger as it . . . becomes more concentrated." When the difference in force becomes equalized, you can shift your defense to offense. Consider the enemy's position, the condition of your reserves, and the terrain advantages when deciding to transition from defense to offense.

Psychological and Leadership Considerations

1. **Defensive Morale and Leadership**: Strong leadership is crucial in maintaining high morale. Defense should be portrayed as a proactive choice rather than a sign of weakness.

2. **Outcome**: The only outcome for defense is to repulse the enemy, rarely to defeat them. Your goal is to retard the enemy and to prevent a

final outcome.

3. **Endgame Strategies**: Know when to cease defensive postures based on strategic assessments of material and moral conditions. This may involve transitioning to peace negotiations or preparing for a counter-offensive.

4. **Victory**: Victory goes to the side that most spares their forces and who best understands how to make their forces act at the right moment. More on Victory in the next chapter.

Schalk's Great Maxim for defense: Once the attack commences and the first volley is fired, immediately advance on the attacker.

Clausewitz on Defense

Defense is not usually a negative posture. Just as in American professional football, where the defensive team can win the game, so then is defense a positive force in strategies. Clausewitz emphasizes that defense has several strategic advantages and should not be viewed merely as a passive posture. The proper use of defense can dictate the tempo of conflict and shift the strategic initiative to the defender.

Implementing Clausewitz's Defensive Principles

The following are situations described by Clausewitz where defense is best employed:

     a. **Timing**: When an offense by an opponent is certain and preparation time does not provide you with an opportunity to strike first or to prepare a counter-offense

2. **Morale and Attitude**: When an offense would be politically unacceptable; where you are better off to let the other person strike the first blow

3. **Relative Strength**: When the probability of the success of an offensive is low and when taking your time will provide you with a better opportunity to increase your strength and then to attack at a later time

4. **Physical**: When your position is strongest where you are and when a defensive position provides an advantage over the offender; your position advantage may be due to physical conditions or familiarity with the surroundings

5. **Stalemate**: When full defeat of the opponent is not possible, and the opponent will not be able to defeat you where you are

These scenarios illustrate the strategic wisdom in choosing defense over offense, ensuring that resources are conserved and advantages are maximized until a more opportune moment arises.

9. Defeat

Defense differs significantly from retreat. Retreat is often chosen to prevent further loss of resources and to gain time. Remarkably, a retreat can even serve an offensive purpose. Clausewitz suggests that you can wear down your opponent and stretch their supply lines by retreating. Additionally, you can retreat strategically by drawing your opponent into a more advantageous position, such as retreating to a higher ground.

However, a retreat is primarily a defensive maneuver. It can provide crucial physical or psychological rest, especially after extensive offensive operations, allowing time to rejuvenate your personnel. Maintaining your position and rotating fresh personnel is usually more beneficial if resources permit.

When planning a retreat, it's essential to consider your escape routes carefully. Your opponent may anticipate your direction and set ambushes. Thus, having several alternative paths and securing them during your retreat is crucial.

We have highlighted the positive aspects of defense and both the positive and negative aspects of the retreat. Let's summarize the primary purposes for taking a defensive stance:

- To control time by delaying, prolonging, and causing waiting.
- To move to a position where you have the advantage.

Drawbacks to a defensive position include:

- Lack of mobility.

- Lack of progress toward objectives, such as occupying territory or capturing supplies.

- Potential loss of morale if personnel perceive defense as a sign of weakness.

The fundamental goal of defense is to achieve objectives by conserving resources. This conservation is accomplished by:

- Avoiding engagement.

- Protecting resources, particularly those tempting to your opponent.

- Conserving resources while fending off attacks.

To protect your possessions, keep them out of your opponent's view and reach using distance and barriers. Barriers can be natural, such as terrain, weather, and darkness, or human-made, like vaults, walls, trenches, and pillboxes. General Patton, however, argued that fixed defenses are often ineffective because they can be quickly bypassed, as he demonstrated during his rapid advance across Europe.

Besides protecting resources, stationing defensive forces strategically around your location creates an initial resistance point and a communication network to summon reinforcements quickly. This plan relies on rapid communication and the ability to position forces faster than the opponent can breach the initial defense.

A final consideration for defense and offense is knowing when to cease hostilities and, in the case of defense when to surrender. Every conflict reaches a point where the choice is to quit or die for the cause. The inability to continue offense or defense often leads to making peace, surrendering, or fighting to the death. This decision can stem from the exhaustion of resources or the will to continue. The willingness to die for a cause involves complex factors like belief in principles, loyalty, upbringing, discipline, and commitment. Leaders instill this level of dedication through psychological mechanisms; without it, desertions and easy surrenders may occur.

Clausewitz advises that making peace is generally preferable to surrender, as an "honorable cessation" is more favorable than pure subjection. The decision to shift towards peace depends on the following:

- Material and moral defeat.

- The improbability of victory.

- The unacceptable cost of further losses.

In business, surrender may occur when further investment would only lead to additional losses—a concept known as "not throwing good money after bad."

When facing a decision to quit, one of the most challenging choices is whether or not to expend more resources. In a losing situation, deploying additional resources can sometimes turn failure into success. Achieving success in this manner can overshadow initial losses in records and memories. However, if you waste more resources and fail, you risk being remembered as foolish, imprudent, or cruel.

When deciding whether to quit, consider how your opponent will react if they believe they have won. An opponent who thinks they have secured victory may exhibit a range of behaviors, from intolerable cruelty to predictable arrogance or even understanding and compassion. By studying history and understanding your opponent, you can gauge their likely behavior towards you and your supporters in defeat, empowering you to make an informed decision. This anticipated behavior should be a major factor in your decision to surrender.

To prevent the surrender of drafted troops with questionable resolve, military leaders have often spread rumors about the terrible treatment and torture expected at the hands of the enemy. Fear of these consequences can effectively reduce the incentive to quit.

As we explore defensive strategies, remember the philosophy of Carl von Clausewitz, who strongly believed that the best approach is to begin on defense and end on offense. Other alternatives may offer short-term solutions, but transitioning from defense to offense is often the key to ultimate victory, inspiring and motivating you to keep pushing forward. less advantage in terms of impact on morale.

Consider the following strategic options:

1. **Begin with Defense, End with Offense**: Goal: Exhaust the enemy's resources first before launching a counterattack. This strategy allows you to weaken your opponent while conserving your strength, positioning yourself for a decisive and strategic offensive move when the enemy is most vulnerable.

2. **Begin with Defense, End with Defense**: Goal: A losing strategy unless you can outlast a prolonged siege. This approach focuses on endurance and resource management, hoping that the enemy will deplete their resources and morale before you do. However, it carries a high risk of eventual defeat if the siege cannot be broken, creating a tense and uncertain situation.

3. **Begin with Offense, End with Offense**: Goal: Effective if executed swiftly and relentlessly, leading to a decisive victory. This aggressive strategy aims to overwhelm the opponent with continuous pressure, leaving them no opportunity to regroup or counterattack. The success of this strategy depends on maintaining momentum and not allowing the enemy any respite, creating a high-pressure and intense situation.

4. **Begin with Offense, End with Defense**: Goal: This scenario often results in defeat; it is better to seek negotiation. Starting with an

offense but shifting to defense usually indicates a loss of initiative and momentum. This transition can expose weaknesses and create opportunities for the opponent to exploit. In such cases, negotiating terms may prevent a total defeat.

When you are the underdog, with fewer resources, it is imperative to plan a robust defense to support your offensive actions. Refer to the list of specific defenses in the following chapter for detailed strategies.

In conclusion, if you find yourself on the defensive, you should immediately begin planning your offense. Your goal should be to transition from defense to offense to secure victory. Relying solely on defense without a plan often leads to defeat.

Defeat

Jonathan Evan Maslow identified these reasons for defeat:

1. **Complacency**: When you hear "we've got it in the bag," that is when the momentum is lost. Leaders and unrivaled champions begin to feel invincible just before they lose. Remember the 1979 L.A. Dodgers? Never get complacent. Watch for your opponent to do so.

2. **Broken concentration or lack thereof**: When Time-outs, delays, or other ways to break an opponent's rhythm makes the essential difference in defeating him or her. In the 1947 U.S. Open, in sudden-death playoff, Sam Snead blew a three-foot putt after Lew Worsham asked for a measurement to see whose ball was closest to the hole.

3. **Choking**: When an individual's performance breaks down under his or her own mental pressure. Choking often strikes the young and inexperienced and may occur suddenly or slowly. The Chicago Cubs had a reputation for many years of choking near the end of the baseball season.

In the history of war, where the outcome is usually determined in favor of the combatant with the greatest strength, exceptions do exist.

What factors cause a weaker force to overcome a stronger one? Sometimes the difference has been on the following:

- The element of surprise

- Better resources—such as people or equipment

- Greater preparation—better training

- Greater determination

- Superior tactics, stemming from experience

- Superior leadership and the higher morale that follows

In every case, the difference was something unexpected, better, greater, or superior. If you are the best at everything, then your chances of defeat are reduced. Even if you have been in a defensive posture for a long time, and your resources are approaching collapse of morale, attacking with no further regard for winning or losing can regenerate better morale.

If you have done your best and have lost, then you can ask for no more than a noble end. Prepare for total destruction or surrender as thoroughly as you did your offense or defense. Before you lose control, or while negotiating surrender, make arrangements. Arrange to protect, destroy, or hide some resources. Plan the end as carefully as the beginning. At the end, you should execute the ending plan as best you can under the circumstances. However, the end is not always final!

One other point about failure—when you are wrong, lie! Not really, we are just kidding. Lying rarely works. When you make a mistake, admit it quickly, apologize for it, correct your mistake, and move on. If you find yourself admitting to mistakes often or admitting to big mistakes, prepare to be removed from your leadership role.

Role of the Leader in Defeat

Failure by the leader can be a cause of defeat. Carol Hymowitz of the Wall Street Journal, reporting on a survey done by the Center for Creative Leadership, listed five reasons why managers fail. These reasons are true beyond business, as follows:

- Inability to get along—on many levels—up, down, peers, outsiders

- Failure to adapt—some people tolerate change well, others do not

- Preoccupation with self—not being a team player, losing focus on company goals

- Fear of action—inability to bring issues to a head or to accept the risk inherent in decisions

- Unable to rebound—inability to weather a setback

Positive solutions overcome the reasons for failure. Be team- focused, accept or better yet create change, watch, and meet deadlines, correct errors as soon as they occur, and accept responsibility.

Failure of one person can be opportunity for another. When a failure occurs to someone else, look at how he/she failed. Did they lack sufficient experience, begin with insufficient resources, apply insufficient effort, lack dedication, fail to concentrate their attack, or make an overall blunder in approach, timing, or execution? If you find one or more of these shortcomings and you have the missing strengths, you may consider attacking where they failed.

How to Recover from Defeat

When you are defeated, you lose control. However, many times people have later regained control. Just recall Billy Martin's baseball coaching career or General McArthur promising to return to the Philippines.

Here are ways to return from defeat:

- Do not hold on to negative energy.
- Decide now to forgive yourself for the failure.
- Finish unfinished business, including apologies.
- Ask for help for yourself from allies.
- Ask the victor to help your people.
- Maintain your dignity; your behavior in defeat is how eventually you will be remembered.
- Set a limit on how far you will go.
- Set out to improve the situation.
- Start to work on another challenge.

- Change your environment (go into exile, leave).

You can return from defeat. Exiled emperors have returned to be reinstated to power. Napoleon did so twice. Abraham Lincoln did so several times. Nixon did so twice. Also, you should remember to not wait until defeat is upon you before you seek a refuge. Your original plans should provide for places of refuge in the event of defeat.

Defeat

In defeat, when your offense has clearly failed or further pursuit of your objective would only incur greater losses, you may face the same situations as described above for the victor, but from your perspective as the defeated:

1. **A total rout**: Your forces are dispersed in disorder.

2. **Retreat**: You withdraw in an orderly fashion to a more secure location to allow you time to regroup, consolidate, and reorganize.

3. **Surrender and occupation**: You turn over all of your resources and leave yourself to the mercy of your opponent. If you survive, you may be allowed to return to a passive life and regenerate strength, rebuild forces, and accumulate resources for future retaliation.

4. **Total annihilation**:You and your resources are destroyed.

Defeat—Conclusions

Defeat may leave you an opening for a new offense. Your victor may be weakened from their struggles and now be more vulnerable than ever with their guard down. They may also be disorganized, be spread thin, and have exhausted their resources and supply network. If so and if you can muster enough resources, mount a surprise attack, and a strong enough offense, you may be able to reverse your defeat. Failure a second time will expose you to a more severe defeat and possible annihilation. Good judgment more than speed is essential in reversing a failure.

Always beware an opponent who has failed in the past. Such an opponent has gained experience and may be much stronger in will. Failure is a learning experience. Failure is less severe than total annihilation obviously because you live to fight another day. Learn what you can from failure and move on to your next endeavor.

10. Victory

VICTORY IS THE success of either your offense or your defense. Victory comes from a series of intermediate successes that are links in a chain. Each intermediate victory can slow you down, depending on how you manage them. Napoleon destroyed every village he passed through on his way to Moscow, including the crops, so that his men would not think of trying to return home and depriving his opponent of resources if retreat became necessary.

Victory

Of course, victory for you means defeat for your opponent (or vice versa), except for win-win scenarios, which do entail strategy but have as their purpose, accommodation rather than defeat. Victories happen at intermediate objectives as well as at the final one. The final one is the only victory that resolves matters permanently or at least for a long time.

As the victor, three major outcomes should be anticipated:

Successful Strategy Outcomes
1. Complete Victory
   a. **Total Rout**:

> *Outcome*: You pursue, destroy, or capture all enemy resources, including personnel and equipment. This outcome involves a decisive and overwhelming victory, with your forces dominating, leaving no room for the opponent to recover or regroup.

> *Key Actions*: Capturing enemy soldiers, seizing equipment, and securing all strategic locations.

> *Result*: Full control over the opponent's territory and complete dismantling of their military capabilities.

   b. **Annihilation**:

> *Outcome*: You entirely obliterate your opponent's forces and resources, ensuring they cannot pose any future threat. This goes beyond the immediate battle, aiming to completely destroy the enemy's ability to wage war.

> *Key Actions*: Comprehensive destruction of enemy infrastructure, supply lines, and communication networks.

> *Result*: Eradication of the opponent's

presence and influence, leaving their territory and assets entirely under your control.

2. **Control and Stabilization**:

Outcome: Surrender and Occupation: You compel your opponent to surrender and establish long-term control over their territory through occupation. This action involves military presence and political and administrative measures to integrate the conquered area.

Key Actions: Setting up governance structures, maintaining order, and integrating the local economy.

Result: Stabilized region under your control, with diminished risk of rebellion or counter-attack.

3. **Punitive Actions**:

Outcome: Retaliation: You respond to enemy actions with severe punitive measures, short of destruction. This strategy aims to inflict significant damage and demoralize the opponent, serving as a deterrent against further aggression.

Key Actions: Targeted strikes on crucial enemy assets, harsh treatment of captured forces, and psychological operations to undermine enemy morale.

Result: Deterred opponent, with a

weakened will and capacity to fight, but still existing as a potential future threat.

Strategic Considerations:

1. Resource Management:

   - Efficiently allocate and conserve your resources to maintain prolonged efforts in both offensive and defensive operations.

   - Key Actions: Prioritizing supply lines, optimizing logistics, and ensuring sustainable resource utilization.

   - Result: Prolonged operational capability and readiness for future engagements.

2. Psychological Warfare:

   - Utilize psychological strategies to weaken the enemy's morale and will to fight. This campaign includes propaganda, misinformation, and psychological operations.

   - Key Actions: Disseminating propaganda, leveraging psychological tactics, and exploiting enemy weaknesses.

   - Result: Demoralized opponent and reduced combat effectiveness.

3. Negotiation and Diplomacy:

   - Engage in negotiations to achieve strategic objectives without further bloodshed. This step can be complementary when total

military victory is undesirable or
impractical.

- Key Actions: Diplomatic overtures, peace
  talks, and treaties.

- Result: Achieving strategic goals through
  peaceful means, reducing the cost and
  duration of conflict.

As the victor, if you do not handle victory so
that it is permanent, you may find that you have only
temporarily suppressed resistance. Obviously, total
annihilation is one permanent solution. The other
can be to treat the defeated with honor and grace. If
you treat your opponent well in defeat, they will be
less inclined to seek retaliation. In some cases, as
America did with Japan after World War II, control
can be gradually returned to your opponent, under a
new constitution. On the other hand, if you treat
them harshly and do not destroy them, then they live
to fight another day, with a greater purpose than ever
before.

Morality in Victory

Morality also plays a role in how success and
failure are judged in the court of public opinion. If a
defeated opponent is treated with dignity befitting
current social norms, then outside opinion will judge
the outcome as good and noble. If a defeated
opponent is treated harshly, then outside opinion
will work against the victor in the future and may
create new opponents.

PART II

Encyclopedia of Strategies and Tactics

11. Offensive Strategies

What follows is a list of specific offenses that you might deploy in any of a variety of strategic circumstances. They vary in scope from small-scale to vast undertakings, covering areas such as business, warfare, personal life, and politics. This arrangement, presented in alphabetical order, encourages you to explore a diverse array of ideas and possibilities.

Abalone Approach

Abalones are unique creatures. They are tasty seafood harvested by divers who pry them off underwater rocks along the coast of California and other parts of the world. What makes abalones unusual is their reaction to being pried off: if they sense an imminent attempt, they clamp down harder, making it difficult to dislodge them. However, if surprised when relaxed, they can be easily pried off. People often behave similarly.

Bureaucrats, in particular, resemble abalones in their resistance to change. When they are aware of impending changes, they resist fiercely. If you are not in a position of authority and need to implement significant changes, it is more effective to surprise them. Alternatively, like abalones, you might succeed if you can get bureaucrats relaxed and carefully explain the personal benefits they will gain from the change. Securing the cooperation of bureaucrats is challenging, especially when their job security is tied to maintaining the status quo.

Abandonment

Abandonment is a strategic approach that involves two key actions:

1. Abandoning an objective that your opponent desires to distract and mislead them.

2. Simultaneously, it means moving towards an objective you genuinely want to achieve.

This tactic can create confusion and divert your opponent's focus, giving you the control to pursue your goals with less interference.

Absurdity

Absurdity can be a powerful way to create awareness. By contrasting sharply with existing norms or expectations, absurdity draws attention to a particular area. Many television commercials leverage this technique effectively, such as the famous "Where's the beef?" ™ campaign. Additionally, absurdity can be used to frustrate and disrupt highly logical and orderly individuals.

Accepted Precedents

Strategies can exploit a person's predispositions. When your opponent views their following action or move as predetermined, you can capitalize on their predictability. People often fall into routine behavior out of habit or comfort, which can be leveraged to your advantage.

Acting Out of Character

With careful thought and timing, shifting from a stable, recognized image to an unexpected one can intentionally alter how people perceive you. Consider these strategic shifts:

1. **Gregarious to aloof**: signals alarm

2. **Distant to friendly**: signals increased openness

3. **Civil to uncivil**: signals severe disagreement

4. **Aggressive to polite**: signals increased acceptance

5. **Opinionated to open-minded**: signals increased trust

Such changes in image can convey powerful messages in a negotiation setting, influencing perceptions and reactions significantly.

Agreements

Much like an alliance, an agreement involves less coordination of efforts and more a temporary exchange of specific promises. An agreement entails trading specific actions where fulfillment benefits both parties. For an agreement to be effective, both parties must receive something of value. Additionally, an agreement often includes penalties for nonperformance. An agreement does not truly exist unless there is a mechanism to penalize one party for failing to fulfill their obligations; otherwise, it is more akin to a partnership. The potential penalties may be explicitly stated or merely implied.

Alliances

An alliance involves coordinating efforts between two or more parties toward a common goal. This strategy is especially significant in societies operating above the law or lawless situations where strength is paramount. Alliances are formed when both parties have something to offer and something to gain. A party in need is unlikely to successfully join an alliance if the other party does not need assistance.

Alternate Outs

When asking someone to make a tough decision, your role is to provide options that guide them toward your objective without requiring a direct yes or no answer. This approach reduces the likelihood of strong objections and minimizes the chance of outright rejection, empowering the reader to actively participate in the decision-making process.

Artificial Catastrophes

In this strategy, you exaggerate an otherwise minor problem or situation to create an artificial crisis. This manufactured calamity allows you to focus on areas that might otherwise be overlooked and can make others feel concerned or even guilty if you don't address these areas. The resulting distraction provides a convenient excuse to deflect otherwise reasonable requests. Additionally, the fabricated crisis gives you a rationale to act hastily or irrationally, using the situation's urgency as justification.

Assertiveness

This strategy functions as a form of blockade. Assertiveness involves clearly stating what you want and unwaveringly maintaining that position. The key is continually restating your request without compromise until you achieve your goal. By doing so calmly and with reinforcing reasons for each repetition, you increase the likelihood that the person you ask will eventually acquiesce.

For example, consider returning defective merchandise to a store using assertiveness. If the store refuses to refund your money, repeat your request. If they still say no, ask to speak to the manager. When talking to the manager, repeat your request. Continue this process. You will often reach a point where someone prefers to comply with your demands rather than resist. This technique can succeed even with unreasonable requests, as some people are so uncomfortable with conflict that they will agree just to avoid it. However, airport gate agents are typically an exception to this rule.

Bait and Switch

This strategy involves offering a bargain that initially seems more appealing than it truly is. For example, you might present an offer that is no longer available when the bargain-hunter tries to claim it. Alternatively, the offer may appear less desirable upon closer inspection, causing people to become interested in a different option that is more attractive and profitable for you. While this approach is illegal in retail sales, it is commonly used in non-retail contexts, such as charity fundraising.

Bathing a Bad Idea in a Good Light

This strategy involves wrapping an unassailable concept around a flawed idea, such as presenting a bad policy in a patriotic light. For example, in 2006, President George W. Bush defended his increasingly unpopular Iraq war policy by arguing that we could not abandon a war where so many brave men and women had sacrificed their lives. By framing his stance this way, he implied that opposing his policy would be equivalent to disrespecting the fallen heroes of the country.

Battle

A battle is a struggle between the main forces of two opposing groups. Once initiated, it must continue until one side achieves victory or withdraws. This fundamental strategy is called the "true center of gravity of war." However, battles occur in settings beyond warfare. For example:

1. Intense political competition just before an election

2. The clash of two competing businesses in the marketplace

3. The struggle of factions for control of a company

These non-war examples share similarities with battles in war. A battle is not a static event but a dynamic process consisting of numerous individual struggles. Each participant strives to annihilate or control as many opponent resources as possible, leading to a constant ebb and flow of advantages gained and lost. This dynamic nature of battles keeps the participants engaged as they navigate the ever-changing landscape of conflict.

A battle represents expending resources to achieve an objective, such as bringing an opponent to submission, controlling territory, or eliminating a competitor.

Bear Hug

Inspired by Carl von Clausewitz, this strategy describes a seemingly loving embrace that becomes unbreakable and ultimately "unbearable." For example, in business, a company targeted for takeover might be approached under the guise of a friendly and favorable merger, a diversion tactic that serves to distract from the true intention of a hostile takeover.

Becoming a Shark

This strategy involves exploiting the vulnerability of someone who earlier attacks have weakened. The "smell of blood" is a signal detected only by the most intense aggressors, indicating an opportunity to strike.

Blackmail/Extortion

Extortion is a strategy in which power or position is used to obtain property, funds, or patronage. Blackmail is one form of extortion. It involves extorting money or other value by threatening to expose a criminal act or revealing damaging or embarrassing information. For example, if you threaten to harm your opponent and they agree to "redistribute their wealth to your benefit" to avoid the damage, you use blackmail as your strategy. Blackmail can be illegal or legal, depending on the circumstances.

In business, negotiations sometimes take the form of blackmail. For instance, an executive negotiating the terms of his employment termination might receive additional severance pay in exchange for his silence about specific company affairs, including the agreement itself. In this scenario, the executive and the company engage in mutual blackmail, leveraging their position to achieve a favorable outcome. Blackmail involves identifying and exploiting a vulnerability, often requiring careful observation and investigation to uncover weaknesses, secrets, or fears.

While physical coercion is often easier than

changing minds, blackmail can effectively influence decisions, especially when threats target personal safety or family members. However, minds can also be altered through emotional appeals or enticement, such as bribery or rewards. An example of subtle blackmail might be a department store lowering the air conditioning to make customers gravitate toward coats mistakenly shipped by northern suppliers, thus creating an artificial demand.

It is important to note that violating the law is neither a safe nor ethical course of action.

Blockade

A blockade is formed when a more substantial power obstructs the supply lines of a weaker power to force compliance. It is an alternative to direct war, leveraging political or economic pressures on the vulnerable entity. The more substantial power cuts off critical inputs and outputs to create pressure on the weaker force to meet its demands. For example, the United States has maintained an economic blockade against Cuba for over twenty years. This blockade began during the Cuban Missile Crisis when a direct attack could have escalated into a full-scale war with the Soviet Union, Cuba's ally at the time.

Bluffing - Strength in Weakness

Appearing strong when weak is a bluff that relies on projecting strength despite vulnerability. This tactic is only effective when your opponent believes in your facade and does not challenge you. Therefore, it should be used sparingly. If you are desperate, avoid using this strategy if your opponent knows of your desperation—it will fail. Instead, use it only if your opponent is uncertain about your strength.

An example of this kind of bluffing occurs in poker. You might bet high to convince others you have a strong hand when, in reality, you have a weak one. This strategic aspect of bluffing can keep you engaged in the game. Sometimes, players with strong financial positions bluff just for fun or because they can, but this adds an unnecessary element of risk.

Bluffing - Weakness in Strength

Appearing weak when you are strong is a bluff that relies on projecting an impression of vulnerability despite actual strength. This tactic is successful when your opponent believes in your facade and decides to attack, thinking you are weak.

A real mess can occur when both you and your opponent are bluffing. Additionally, after losing to a bluff, your opponent will often come back stronger to get even, potentially overextending themselves. You can use their overreaction to your advantage, even leading them into a trap.

Box In

When you box in your opponent, you surround them. Concentrating your forces and encircling the opponent eliminates their escape routes and cuts off their supply lines. Done thoroughly, you may be able to just wait for them to run out of sustaining materials.

Braking Action

This timing strategy involves slowing down the action. This approach can be helpful in several situations:

1. **Need for Preparation**: When you require more time to be fully prepared, slowing the action allows you to get ready. However, be cautious that the additional time benefits your opponent as much as it does you.

2. **Distraction Tactic**: When your opponent is in a situation where a few individuals can keep a large number occupied for an extended period, slowing the action distracts them from their primary objectives.

3. **Wearing Down Superior Numbers**: When facing an opponent with superior numbers, slowing the action can help wear them down over time.

Breaking up a Team

When a team shows signs of decreasing effectiveness, consider breaking it up to identify the driving forces and followers among the members.

Br'er Rabbit

This strategy is inspired by the children's story of Br'er Rabbit. In the tale, a rabbit tries to escape from a fox by using reverse psychology, pleading, "Please don't throw me in the briar patch." Predictably, the fox throws him into the briar patch, where the rabbit is safe and secure. This strategy involves using reverse psychology to achieve your desired outcome and works best when dealing with someone who tends to do the opposite of what is suggested.

Brinkmanship

This strategy involves deliberately creating a recognized risk to maintain pressure on an opponent. A simple example is: "Don't move, or I'll shoot." This approach effectively halts any further encroachment and serves as a robust counter to a "nickels and dimes" strategy, a tactic where your opponent makes slow but steady progress through small, incremental successes. This strategy assumes that your opponent will respond to your threat as intended.

Broken Wing

In this strategy, you appear weak while actually being strong. This approach is a powerful way to divert your opponent from a critical location, issue, or your true weakness. Your actions should be perceived as weaknesses but should conceal your underlying strengths. It's crucial to understand that your opponent's perception of your weakness is what allows you to exploit their vulnerability and surprise them.

Consider the example of a lowly sparrow. If you approach a fledgling that has fallen from its nest, its mother will land nearby and feign a broken wing to draw you away from her baby. She will fly off once you are far enough away, having successfully diverted your attention. Fortunately, sparrows cannot carry or use guns.

Bullying

Bullying involves asserting dominance to compel the other person to do your bidding. This strategy forces action by leveraging your strength. For it to succeed, the bullied person must submit to your dominance. If they resist, your strength must be sufficient to overpower them; otherwise, the situation can backfire, and the tables may be turned.

By-product

A by-product of an action occurs when you get more or different results than expected. For example, if you embarrass someone to discredit them and they resign, this is a by-product. It's crucial to consider the potential extreme outcomes of your strategies. Avoid actions so severe that your opponent may take an unexpected and extreme response, such as suicide.

Call-In

The name of this strategy comes from duck hunting. A hunter uses a reed device to mimic the sound of a duck, attracting other ducks to the location and making them easy prey. Similarly, this approach can lure your opponent within striking range. The idea is to get your opponent to approach you rather than pursue them.

Chaos and Noise

The use of chaos and noise can be an effective distraction. According to Stan Winnie, a business associate, his high school baseball coach used this strategy successfully during likely bunting situations with one out and a runner on first base. The coach instructed every team player, including those in the dugout, to start yelling while the first and third basemen charged toward home plate. This commotion often caused the pitcher to turn and throw to second base, catching the runner between first and second.

Pickpockets use a similar strategy, creating distractions through noise and activity to divert attention while they steal.

Chemical Attack

The use of chemical warfare, which involves deploying toxic chemical substances as weapons, is widely considered abhorrent. Despite its universally condemned nature, this strategy has been threatened by terrorists and employed by malevolent regimes, such as that of Saddam Hussein.

Cherry Pick

This strategy involves selecting the best options, like a miner extracting valuable gems from a mine. By cherry-picking the most precious stones, the miner leaves behind the less valuable ones. This selection process deviates from the usual order and instead follows a sequence that offers immediate advantage. Commissioned salespeople often focus on items that provide the highest payoff or are easiest to sell rather than following the sales manager's intended priorities. This strategy is inherently one-sided, benefiting the "picker" while disadvantaging the "resource provider."

Close-in Fighting

Engaging resources from a long distance is generally ineffective, as it allows your opponent to incur fewer losses while you expend resources without gaining any ground. Close-quarters engagement is essential to force your opponent to retreat in disorder. Only at close range can you effectively disrupt and overpower your opponent.

Closeting

When you closet someone, you isolate them to focus solely on what you have to say. This strategy is often used in negotiations where privacy and avoiding public scrutiny are crucial. In the U.S. Congress, closeting frequently occurs in hallways and during dinners.

Coalition Surprise

Coalitions are powerful tools in any conflict. Discovering that one is facing a coalition can have a devastating psychological impact, significantly damaging the morale of the opponent's forces. By forming a coalition in secret and revealing its existence at a strategically opportune moment, you can leverage the element of surprise to deliver a potentially overwhelming blow to your opponent.

Collusion

In this strategy, you join with an ally to act together against an opponent. This strategy is particularly effective in financial cheating. Enron is a recent example. In auditing, procedures are most effective at catching single cheaters. When two or more people join to cover up an improper act, catching the culprits becomes much more difficult. Complicity is an essential element of this strategy. Watch for a double-cross, being framed, and blackmail.

Come In Over

To bring pressure to a negotiation that isn't going well, escalate the situation by having your superior approach the superior of the person you're negotiating with. This action aims to secure a better deal at a higher level. Large consulting firms effectively use this strategy by assigning managing partners, partners, project managers, and consultants to engage with different levels within the client company.

Common Enemy

Create a shared enemy to unite yourself and a potential ally. By establishing a common adversary, you generate the motivation your neighbor needs to become your ally. This strategy is frequently employed in business to form strategic alliances.

Communications that Lead

A leading communication is a simple, understated, crafted message that prompts the reader to ask a specific question. This technique allows you to guide the reader into inquiring about the very issue you want to address but cannot bring up directly. In other words, you subtly lead the reader to ask the question you want to answer.

Company Politics
Classifications

Company politics can be classified by their purpose:

1. **Personal Attacks**: These arise from personality clashes, disagreements, or other personal reasons. Initially, try to connect personally with the other person. If that fails, consider the person an adversary.

2. **Turf Battles**: These involve struggles over

the type of work, areas of control, the number of people supervised, or sources of power such as influence or position.

3. **Competing for Priority**: This involves fighting to prioritize your needs over those of others.

4. **Competing for Promotions or Power**: A more significant form of turf battle at higher levels of the organization. Each new promotion offers increasing degrees of power and benefits, with power growing exponentially as you move up the chain of command.

Surviving Company Politics

If you find yourself caught up in company politics, there are many techniques you can use to mitigate the impact:

1. Act as if you already know the facts.

2. Guess aloud: For example, "Looks like there's going to be a reorganization."

3. Associate yourself with the leaders.

4. Ask for information.

5. Avoid answering questions directly that force you into someone else's framed responses (e.g., "If ____ happens, then what will you do?" "Are you going to ____ or ____?" "Which is it, ____ or ____?")

6. Be sure of your facts.

7. Put positions in writing only when necessary:

    - To establish ownership of your idea.

    - To praise people.

    - To document your disagreement.

    - To keep your boss informed.

    - To summarize positions if presenting a balanced view. Do not document controversial ideas, irresolvable issues, disagreements, or angry words.

8. Require explanations. Ask "why" often.

9. Challenge inappropriate comments -- "Why do you say that?"

10. Use admission/deflection rather than being defensive: For example, "Gee, I am afraid that spoiled shipment was my fault. I neglected to double-check Joe's inspections. Joe, you did say that you inspected it, didn't you?"

11. Build alliances, especially with your boss.

12. Avoid alliance traps:

    - Relying on an alliance before it is formed.

    - Asking for support without knowing your ally's plans.

    - Overusing an alliance.

    - Tying your request for support to other people.

    - Letting alliances lapse.

13. Watch your body language and observe others. Pace your body language with theirs.

14. Maintain a positive, supportive, and team-oriented attitude internally (your body language will follow).

15. Use your voice and speaking skills to focus thoughts clearly and support your positive attitude.

16. Walk, talk, and act as if you owned the whole company.

Compromise

This strategy involves seeking a middle ground by compromising on some of your demands in exchange for having others granted by your opponent. The key is understanding what is important to your opponent and what is not and knowing what you are willing to concede and what you are not. This mutual understanding is essential for effective conflict resolution.

Communications

Communications are most effective when they include the following elements:

1. **Defensible and Accurate Evidence**: Present true, accurate, and complete evidence.

2. **Objective Delivery**: Deliver the evidence without adjectives, judgment, bias, or opinion.

3. **Relevance and Clarity**: Eliminate minor and peripheral facts, removing insignificant details.

Consistently following these guidelines builds a track record of credibility and reliability. This approach will steadily expand your audience, increasing your impact and effectiveness.

Computer Manipulation and Cyber Disruption

You can manipulate information in various ways if you control the computer systems. For example, you could:

1. **Review Reports**: Examine every report before it's issued and use the information for your purposes.

2. **Restrict Progress**: Delay the delivery of computer outputs to hinder another's progress.

3. **Selective Reporting**: Produce reports

selectively casting a negative light on an opponent.

In addition to these tactics, governments and other entities often engage in hacking and cyber disruption to further their agendas. This interference can involve:

1. **Data Breaches**: Unauthorized access to sensitive information to gain a strategic advantage.

2. **Denial of Service (DoS) Attacks**: Disrupting services to incapacitate an opponent's operations.

3. **Misinformation Campaigns**: Spreading false information to manipulate public perception or destabilize an adversary.

These tactics highlight the significant power and potential for misuse inherent in controlling computer systems and cyber capabilities.

Concentration

If you can concentrate your enemy in one confined area, your superior forces can severely damage their resources. Concentrating the enemy maximizes the chances of beneficial outcomes, increases accuracy, and enhances the likelihood of hitting multiple targets with each action.

Consider the example of shooting a shotgun at different sets of clay pigeons:

- **One clay pigeon**: Shooting at a single clay pigeon demands precision. You may hit or miss, but you can only hit one target at best.

- **Widely dispersed set of multiple clay pigeons**: When the pigeons are spread out, your odds of hitting a target increase slightly, as missing one might result in hitting another.

- **Tightly grouped set of multiple clay pigeons**: When the pigeons are closely clustered, they present a larger collective target. Targeting a tight group enhances your likelihood of hitting one or several with a single shot.

If multiple shooters are focused on a tightly grouped set of targets, the potential for substantial damage increases significantly. This concentration of force makes any attack more effective.

Constant Threats

This strategy involves making a series of smaller threats that collectively create significant pressure on your opponent, resulting in a greater impact than a single large threat. The most effective counterstrategy is to ignore these incremental threats.

Conveyance of Evidence

Complete and effective communication must include a commitment, promise, allegation, statement, and supporting evidence. Weak evidence can be made convincing through careful structuring and presentation. Providing three reasons, even if they are weak, can significantly enhance the perceived strength of the argument. <u>Presenting three points</u> has a psychological impact that often works to persuade others, even with weak points.

Cop a Plea

This strategy involves negotiating a settlement at a lower cost than you might otherwise incur by trading off uncertainty against expense. A typical example is the plea bargain, where the defendant agrees to a lesser charge or penalty to avoid the uncertainty of a trial.

Coverall

This strategy involves presenting a weak reason for an action, which, when attacked, allows you to unveil a more solid, underlying reason. This approach often silences any further questions. For example, a company might justify a policy change by citing minor cost savings. When this reason is challenged, the company can reveal the true motivation to comply with new regulations, which is a much stronger justification.

Covert Actions

This strategy is a more formal and severe version of a Dirty Trick (see below), where a political power uses its resources and influence to:

1. Undertake an action, such as an assassination.

2. Disguise the source of the action to conceal its involvement.

While this strategy is more common in lawless societies, it can sometimes be deemed lawful or justified in specific contexts or societies.

Creating Contrast

This strategy involves identifying negative characteristics or factors about your opponent and contrasting them with your positives when competing for the same resources. For example, negative campaign advertisements employ this tactic by highlighting every possible adverse decision an opponent has made and contrasting them with the candidate's positive choices to enhance their image. This approach focuses on tearing down an opponent to make oneself appear better. It is commonly used in company politics, abusive personal relationships, and other situations where competitive, unethical, and power-hungry individuals vie for control.

Creating Panic

This strategy involves exploiting a situation where a crowd's access to escape routes is constrained. This trap can lead to a complete rout of the enemy and a panicked retreat in war. However, we strongly advise against using this strategy with innocent civilians and even in wartime. Yelling "fire" in a crowded theater is illegal for good reason. Incidents such as fires started by fireworks in clubs, suicide bombers in nightclubs, and riots at soccer games have all resulted in horrific deaths due to crushing. Such tactics cause unnecessary harm to innocent people and should never be employed.

Cross and Double-Cross

The cross involves sending a person to spy on your opponent. The double-cross occurs when your opponent turns your spy against you. The triple-cross is an even more intricate variation, where you secretly join forces with your opponent to betray your supposed ally.

Cross Over

This strategy involves shifting a problem from one area to another. For example, if you are over budget, you can transfer some costs to another under-budget organization. This approach can benefit both parties: the first department meets its budget, and the second does not have to justify its expenditures to maintain or increase its budget for the next cycle. In extreme cases, you might even transfer costs to another department without informing them or testing whether they notice the discrepancy.

Cut Off at the Pass

This strategy is a variant of the Preemptive Decision (see below). For example, consider a committee whose assignment overlaps with your area of responsibility and whose conclusions conflict with your objectives. You ask to be kept informed of their progress. By staying informed about their discussions, you can stifle their initiative by making the decision yourself before they can make their recommendations.

This strategy involves knowing and using a shortcut. You take a shortcut to reach a goal before someone else, positioning yourself to prevent your opponents from achieving their objectives. For example, you could prevent a building developer from constructing in an undesirable location by convincing the zoning board to reject their application.

Cut Off the Head

To eliminate a poisonous snake, you can cut off its head. The same strategy can apply to an opposing force. By destroying their leadership and ability to lead, you undermine those who followed that leadership, disrupting their direction and possibly their will to fight.

Cut Off with a Lawsuit

A tire company collaborated with a supplier on developing a new technology for several years, ultimately gaining a significant advantage in the marketplace. However, the tire company failed to secure an exclusive agreement with the supplier. As a result, the supplier began negotiating a similar deal with the tire company's competitor. Faced with the prospect of losing its competitive edge after making a substantial investment, the tire company took decisive action. They sued the supplier for nonperformance of their original contract. Upon hearing about the lawsuit, the competitor backed off and did not finalize the agreement with the supplier.

Cutting Off the Angle of Attack

An attack originates from specific directions and can be blunted by countering one or more of those directions from an angle that deflects the attack.

Cutting Off Their Information

This strategy highlights the crucial role of information. Imagine someone sneaking into your office or home and removing all files, manuals, papers, phone lists, and essential documents. If your office is emptied, you will be unable to function effectively for days. This concept extends to information on telephone lines or computers, showing the devastating impact of losing critical data.

Cutting Off Their Water

This strategy comes from the early Western feuds between the cattle ranchers and the sheep ranchers, when a rancher upstream would divert the flow of water to drive the other rancher away. In an offense it means to stop the flow of someone's key resources. Cutting off your opponent's supply of resources will damage his or her plans.

Dagger Thrust to the Heart (Gravity Center)

The most effective move against the opposition is concentrating your force on their center of gravity. This direct frontal assault targets the core of their strength, presenting a clear risk of either failure or success with little middle ground. Such a strike effectively destroys resources and morale but requires superior resources to succeed.

Damage

Damaging your opponent's resources weakens their strength and effectiveness and enhances your chances of success. One method is sabotage, where damage is inflicted secretly, aiming to achieve success without being identified as the perpetrator. Another approach is direct and open destruction. In business, this could involve damaging a competitor's reputation. In politics, it might include smearing an opponent (see Smear Campaign below).

Deception

With this strategy, you resort to trickery or cheating to deceive your opponent. Fraud, counterfeiting, faking, hoaxes, shams, and swindling are all forms of deception. This approach involves fooling your opponent through unfair, illegal, or surprising methods considered unjust by some standards. It works because your opponent does not anticipate you operating outside the usual bounds of laws, rules, or norms. Deceit is most effective when the risk and penalty for breaking the rules are minimal or when success is almost guaranteed.

Decision-Splitting

When faced with an unfavorable decision, split it into parts, ensuring that the first part receives a favorable response. A positive response for the first part sets a positive tone for the subsequent parts. You can address the remaining parts later when you have more time to gather information and support.

For example, you want a program approved, but your boss insists on a budget reduction to compensate for its cost. If your budget doesn't allow for this, agree to start the program and promise to return in two weeks with the details of the budget reduction. When the time comes, find reasons to delay providing these details. Repeat the delays as long as possible, hoping the requirement might be forgotten. Meanwhile, continue highlighting the benefits of the project. Once the project is well underway, the budget reduction might be overlooked. If the issue resurfaces, present an inadequate budget reduction and be prepared to be sent back to revise it.

Decoys

This strategy is diversionary: you draw attention to a target other than the one you aim for. For example, you might tell one person something while ensuring the intended recipient overhears the message.

Decoy Doubled

The double decoy strategy involves using deception to mislead opponents about your true intentions by creating multiple layers of diversion.

For an example, consider Harold Simmons. In March 1990, Simmons attempted to gain control of Lockheed. On March 26, 1990, Lockheed publicly appealed to their shareholders in the Wall Street Journal, urging them to reject Simmons's offer. They cited a quote from Simmons in the October 19, 1989 edition, where he claimed, "Lockheed is just a decoy... I have several decoys out there." This comment was intended to suggest that Simmons was not serious about his offer.

However, a few weeks later, it became clear that Lockheed was Simmons's original target. His public comment about Lockheed being a decoy was itself a decoy. He performed a "double decoy," using the concept of decoys to mask his real intentions.

For another example, consider Harold Simmons. In March 1990, Simmons attempted to gain control of Lockheed. On March 26, 1990, Lockheed publicly appealed to their shareholders in the Wall Street Journal, urging them to reject Simmons's offer. They cited a quote from Simmons in the October 19, 1989 edition, where he claimed, "Lockheed is just a decoy... I have several decoys out there." This comment was intended to suggest that Simmons was not serious about his offer.

However, a few weeks later, it became clear that Lockheed was Simmons's original target. His public comment about Lockheed being a decoy was itself a decoy. He performed a "double decoy," using the concept of decoys to mask his real intentions.

For a recent example, consider the marketing strategies employed by companies like Apple and The Economist. Apple often uses decoy pricing to nudge customers toward mid-range products. They present multiple models with different price points, such as the standard iPhone, the iPhone Pro, and the iPhone Pro Max. The Pro Max is priced significantly higher, making the standard iPhone and the Pro model appear more reasonably priced. This tactic subtly influences consumers to choose the mid-priced option, which seems like the best value (SaaS Genius, Shopify).

Similarly, The Economist uses a decoy subscription model to drive sales. They offer a digital-only and combined print-and-digital subscription at the same price. By doing this, they steer customers towards the combined subscription, which appears to offer more value for the same cost, thereby increasing their overall sales.

These examples illustrate how the double decoy strategy can be effectively used in business to manipulate consumer choices and enhance profitability.

Key Points:

- Decoy Pricing: Using multiple product options makes a target product appear more attractive.

- Influence on Consumer Behavior: Leveraging psychological biases to guide decision-making.

- Applications: Common in tech and media subscriptions to maximize perceived value and sales.

These strategies highlight the sophisticated use of deception in modern marketing, drawing parallels to the double decoy strategy in more traditional competitive scenarios.

Delay Strategy

This strategy focuses on controlling timing, a common approach in business and political environments. When faced with an imminent decision you wish to avoid, raise a confusing issue. Once the problem is introduced, request more time to resolve it. Arguments like "we haven't done all the due diligence that we should" often suffice to gain a delay. Most decisions do not need immediate action; additional time is usually granted for any logical reason. Once extra time is secured, you can either work on undermining the decision or seek further delays.

This approach allows you to manage the decision-making process effectively by leveraging time to your advantage.

Delivering Bad News

Avoid being the bearer of bad news, as the recipient will associate you with the negativity. Conversely, whenever possible, deliver good news to be favorably associated. The preferred approach is to be present when someone else delivers the bad news. An even safer strategy is to ensure the bad news is delivered before you arrive, allowing you to be the first to empathize with the recipient. This approach maintains your trustworthiness while leaving the negative impression to someone else.

Destroying Trust

Trust is the foundation for stability and peace. When trust is disrupted, any balanced situation becomes unstable. For example, peace efforts in the Middle East have frequently been undermined by attacks from all sides that destroy trust, perpetuate conflict, and prevent lasting peace.

Destroying trust can be achieved by creating the perception of deception and betrayal. Deception involves deliberately misleading others, eroding their confidence in your honesty. Betrayal, or the perception of betrayal, on the other hand, consists of breaking promises or confidences, profoundly damaging the trust that has been established.

Maintaining and building trust increases the chances for stability and peace.

Die Fighting or Fight to the Death

When you are too weak to defend, attack. Fight with the determination of someone who has nothing to lose. The willingness to die for your cause instills a sense of fearlessness. During World War II, Japanese soldiers exemplified this strategy by cutting an artery under their arm before charging, making them feel "already dead" and eliminating their fear. This act represents the ultimate strategy of turning perceived weakness into an aggressive, fearless offense.

Direct Aggression

Direct aggression can manifest in four stages:

1. **Limited Engagement**: This involves a selective or partial deployment of forces. For instance, one Mafia mob might try to encroach on another's territory through targeted assassinations.

2. **Total Engagement**: This is the active state of offense, where all available forces are committed to the conflict.

3. **Disengagement**: This stage involves a formal or informal cease-fire or the end of hostilities.

4. **Disarmament**: This stage occurs when opposing parties agree to remove or destroy

their weapons, either through negotiation or after the conflict is resolved. by force of the means of further attacks

Dirty Tricks

Nixon's political organizers (and Donald Trump's) excelled at this strategy. Dirty tricks involve actions designed to embarrass, discredit, or make a fool of an opponent, typically without the perpetrator being linked to the outcome. The elements of a dirty trick include:

1. **The Setup**: Arranging all necessary elements for the trick.

2. **The Trigger**: An action by the opponent that initiates the consequences.

The impact of the results depends on the intricacy of the setup. This strategy is a mild form of deception, baiting, and other tactics that lure the enemy into a trap.

Disinformation Campaign

Similar to a smear campaign, this strategy involves distributing false information through credible channels to legitimize the falsehoods. A notable example is the effort by American Express in 1990 to discredit Edmond Safra. As the Wall Street Journal reported on September 24, 1990, American Express chairman James Robinson allegedly used private detectives and other operatives to spread false information about Safra through various news sources.

A more recent example involves the 2016 U.S. presidential election, where false information was disseminated through social media to influence public opinion and discredit opponents. This tactic was employed by various groups, including state-sponsored actors, to create confusion and distrust among voters.

This strategy involves intentionally providing erroneous information to deceive your opponent and can also be used to test the loyalty of subordinates. By leaking false information to someone suspected of disloyalty, their betrayal can be uncovered if they relay the falsehood to your opponent.

This strategy is most effective when:

1. You are sure your opponent will publicize the false information.

2. You can easily disprove the false information after it has been made public.

If the disinformation is accepted but not used, your effort is blunted. Be wary of information that seems too good to be true, especially from unknown sources or those suspected to be aligned with your opponent.

Displays of Strength

In this strategy, you regularly display your strength. You can do this through direct demonstrations, assertive body language, expanding your territory, or controlling the flow of resources and information. Each method clearly signals dominance and capability, reinforcing your position and deterring potential challengers.

Distraction

A minor attack can strategically pin your opponent down to one location or divert them from their intended path. The objective is to divert resources, mislead your opponent, or slow their advance rate. If the distraction fails and your intent becomes obvious, your opponent's reaction will likely be substantial, potentially spurring them to advance more quickly to compensate for the lost time.

Divergence of Timing

This strategy involves carefully staggering the timing of attacks across multiple fronts. By initiating the first wave of attacks, you effectively distract and weaken the opposition, reducing their ability to mount a strong defense against subsequent assaults. This method diminishes resistance for later attacks and creates opportunities to set up strategic ambushes, catching the opposition off guard.

Diversion

A diversion is a strategic maneuver designed to draw an opponent away from their main objective or your main objective. Typically, a diversion begins by attacking a target other than the primary one. The aim of a diversion isn't to capture an objective but to create an opening for later capturing that objective or simultaneously accomplishing another objective.

This strategy might involve dispatching forces to an under-defended location, anticipating the opponent to follow. It could also be a flanking attack on a passing opponent. Another variation is a real attack that appears to be a full-scale assault but is actually a cover for the main attack occurring elsewhere. General George S. Patton famously described this strategy as "hold 'em by the nose and kick 'em in the ass." The diversion should compel your opponent to set up a full defense in the wrong place, utilizing resources needed elsewhere and causing psychological disruption.

A historical example is the lead-up to the Normandy invasion during World War II. General Patton was recalled to England and stationed in an area under German surveillance as part of a diversion. Facades resembling tanks and trucks were even constructed to enhance the illusion.

At a minimum, a diversionary attack can make your opponent appear ignorant and foolish. This type of strategy requires speed, daring, and meticulous planning to appear genuine. Additionally, it's crucial to provide cover for the retreat of the diversionary troops.

However, the outcome of this strategy can be unpredictable. Your opponent might effectively handle the diversion without losing focus on their main objective. Worse, they might ignore the diversion entirely and concentrate their forces on the areas left vulnerable by the diversionary forces.

In a different context, consider the clever use of a diversion by a high school basketball player during a critical game. With only seconds left and his team needing two points to win, the player got down on all fours and barked like a dog while his teammate had the ball. The defending players were momentarily distracted by the barking, allowing the teammate to make the winning shot at the last moment.

Divide and Conquer

This strategy is fundamental. By splitting your opponent's forces, you effectively reduce their protection proportionally.

Domino Effect

This strategy is aptly named after the cascading effect of falling dominoes when they are stood on end in a row, and the first one is toppled. It describes how the fall of one territory to an attacker can lead to the subsequent fall of adjacent territories. In the 1960s, the U.S. Administration was particularly concerned about Southeast Asia succumbing to communism under this strategy. This approach can be effective in wars of attrition.

Draw-In/Trap

In this strategy, you lure your opponent with bait or cunning tactics, drawing them into a vulnerable position where you can engage and defeat them. Typically, this strategy is a one-time setup, but it can also involve a series of efforts to embarrass, ridicule, or otherwise undermine your opponent's integrity. If you can execute a sequence of two or three escalating traps, your opponent may become overwhelmed.

For example, consider the case of two business rivals. One was trying to justify a new facility based on external sales, while the other criticized this justification as being too risky. The trap sprung when the first rival provided additional justification based entirely on internal sales, doubling the needed justification. The trap succeeded because the attacker was eager to embarrass the first rival.

This strategy is a form of diversion. Distracted by the opportunity to "nail" his opponent, the attacker failed to anticipate an additional defense lurking behind the initial justification. As a result, the attacker will likely think twice before targeting that opponent again.

Drone Attack

This strategy employs automated remote control of a robot to attack without any physical danger to the operator. Modern warfare has seen a significant rise in the use of drones and unmanned aerial vehicles (UAVs) that can perform various tasks, including surveillance and combat missions.

For instance, the Israelis used this technique effectively during one of their conflicts. Facing an Arab nation with a substantial missile arsenal, they deployed a large number of drones into enemy territory. This strategy provoked the Arabs to fire off and waste many of their missiles targeting the drones. Once the Arab missile supply was significantly depleted, the Israelis launched their actual attack, with their aircraft facing minimal missile threat.

In a business context, your "drone" could be an unsuspecting colleague whom you send in ahead of you to absorb criticism or blame, thereby protecting yourself and your position. This tactic can divert negative attention from you and allow you to advance your objectives with reduced resistance.

Echelon

Echelon is a parts-stocking strategy in which the selection of parts to be kept on hand depends on various factors, including:

1. **Criticality of the part**: The cost and impact of not having the part available when needed.

2. **Failure rate**: The frequency at which the part fails and needs replacement.

3. **Parts cost**: Cost is a crucial factor in the Echelon strategy. It includes the expense of the part itself, the cost of holding inventory, and the financial implications of stocking parts that may not be immediately used, affecting the overall inventory and financial implications.

4. **Time and cost of transportation**: The duration and expense involved in transporting parts from the source point to the location where they are needed.

Economic Sanctions

Economic sanctions are a powerful tool used to pressure opponents into compliance or to deter actions contrary to one's interests. By restricting access to funds or blockading trade routes, you can influence and modify the behavior of the target nation.

For instance, during the ongoing Ukrainian invasion by Russia, various countries and international bodies have imposed extensive economic sanctions on Russia. These sanctions target critical sectors of the Russian economy, including finance, energy, and technology, aiming to undermine their economic stability and force a reevaluation of their aggressive actions.

A straightforward example of sanctions at work is a parent controlling an older teenager's behavior by withholding their allowance. On a larger scale, the economic sanctions placed on Cuba by the U.S. government illustrate how prolonged economic pressure can be used to push for political change.

In the case of Russia, the sanctions have included freezing assets, restricting access to international financial systems, and banning the export of crucial technologies. These measures aim to weaken Russia's economic capabilities and deter further aggression by making the costs of continued conflict unsustainable.

Empire Building

This strategy is used to build a power base. In business, it involves accumulating departments and hiring additional personnel. It is increasing the resources under your control to enhance your power. A crucial element of this strategy is that another person holds the ultimate authority to allocate these resources. Your power grows when this person chooses to allocate more resources to you, especially when competing with others for the same resources. The key to securing more resources is "trust." When seeking increases, ensure that every proposal is based on a solid, unshakable foundation to build and maintain trust steadily.

Escalation

This strategy begins with justifying and obtaining approval for a small-scale version of the program you desire. Gradually, you expand the program until you achieve the full scope of your original plan. For example, you propose initially developing a single training module and then expand from that module to a comprehensive training school.

Expressing Pain

A negotiating strategy where you react with visible discomfort to a counteroffer, express frustration, leave the room to regain composure, and then return still visibly upset to signal that you have reached your limit.

Extortion (See Blackmail above)

False Identity

Establishing a false identity can grant access to places you might otherwise be prohibited from entering. By assuming your opponent's identity and subsequently breaking the rules, you can potentially cause penalties or consequences to be attributed to them instead.

False Justification

When an action cannot be justified on its own merits, you may attempt to rationalize it using other unrelated yet more noble-sounding reasons. To increase the likelihood of success, provide at least three reasons, even if their soundness is questionable.

False Storm

A false storm is a two-person negotiating strategy where one partner creates a staged outburst to influence the opponent. As the opponent enters the room, one partner pretends to be enraged about the contents of a folder, making harsh comments about the prices listed and criticizing suppliers for their exorbitant charges. The enraged partner then leaves the room with the folder, leaving the other partner to conduct the price negotiation.

This tactic allows you to send indirect messages to the opponent without directly confronting them, setting the stage for more favorable negotiation terms.

Feigning Disorganization

Appearing disorganized can lead your opponent to mistakenly conclude that you are unprepared and not ready for confrontation. Much like a feint, this tactic can lure your opponent into attacking under the false impression that you are vulnerable and unable to defend yourself.

Feints

A feint is a minor attack designed to appear as a more significant threat, unnecessarily causing your opponent to place their resources on high alert. Feints increase your unpredictability and reduce your risks. The key to successful feints lies in understanding what your opponent is focused on and using that knowledge to mislead or misdirect them. This strategy works best against opponents prone to quick reactions or overreactions. For example, an experienced professional sports player might use a feint against an overzealous rookie to exploit their tendency to overreact.

One purpose of feinting is to wear down your opponent and create openings for yourself. Your opponent will typically respond by gearing up and launching a significant attack or counter-attack. As Clausewitz describes, a feint signals an attack, prompting your opponent to prepare for battle fully. Once they are prepared, you can avoid the battle, essentially saying, "Well, maybe another time." This comment has a psychological impact, making your opponent feel used and abused, potentially provoking anger and a subsequent attack.

Feints are particularly effective if you can cause your opponent to peak psychologically several times before launching an attack. During the Vietnam War, the North Vietnamese would attack a U.S. Army camp for several nights in a row before launching a major offensive. When executing a feint, it's crucial to give the appearance of being fully prepared for battle without actually being so. This illusion of readiness, both psychologically and physically, is an essential aspect of this strategy.

However, avoid overusing feints unless you have a significant physical advantage. Repeating any strategy too often can lead to predictability and potential traps. Practice is essential before employing a feint to ensure its effectiveness.

Finding the Mother Lode

The mother lode refers to any recurring source of wealth. For instance, imagine a contractor new to the business and receiving a significant opportunity from a large developer. This developer, capable of providing many future jobs, allows the contractor to prove themselves with an initial project. The contractor is highly motivated to deliver excellent work and hopes to secure future business from the developer. This arrangement benefits both parties: the contractor gains a valuable, recurring source of income, while the developer ensures high-quality work due to the contractor's eagerness to impress and secure ongoing opportunities.

Flattery

Flattery is a compliment with a purpose. The compliment is intended to obligate the receiver to some form of reciprocal action. Compliments also may permit you to take advantage of a person by using their need for positive strokes.

There have been suggestions and reports implying that Vladimir Putin has used flattery to manipulate Donald Trump. According to ex-spies and analysts, Putin has been known to employ "fulsome praise" to appeal to Trump's ego and gain a strategic advantage. For example, during their meetings, Putin often praised Trump's actions and leadership, which some experts believe was a tactic to exploit Trump's susceptibility to flattery and to influence his decisions (The Independent)

Flattening a Tire / Buying Time / Submarining

When you cannot directly oppose an adversary's plan, you may be able to hinder or derail it by employing these tactics:

1. **Tabling the Plan**: Postpone the discussion or decision indefinitely.

2. **Committee Sabotage**: Fill the committee with individuals who lack the necessary competence.

3. **Unrealistic Deadlines**: Set deadlines that are impossible to meet.

4. **Assumption Scrutiny**: Challenge the plan's fundamental assumptions.

5. **Demanding Details**: Request extensive additional details and documentation.

6. **Extended Review Time**: Ask for more time to review the proposal thoroughly.

7. **External Consultation**: Consult your staff or others who are not currently present.

8. **Covert Actions**: Engage in other subtle, behind-the-scenes activities to undermine the plan.

Flush Out

This strategy involves coaxing someone into revealing information or plans they are reluctant to disclose. A classic example from movies is the private detective who, unable to prove a criminal's guilt, strategically closes in on them. Feeling the pressure, the criminal panics and flees, thereby confirming their guilt. This approach hinges on identifying and exploiting a "soft spot."

To flush out your opponent, you might:

- Have the right person ask the right question.
- Make them believe their last escape route is about to be blocked.

Remember, sometimes, just creating enough tension and suspense can be enough to coax them into revealing themselves.

Form a Gang

You can build strength by rallying others inclined to join a cohesive "gang." A gang typically consists of troublemakers led by a charismatic leader with a rebellious mission. By forming such a group, you can leverage the loyalty of its members to extract "taxes" from the local citizenry, a form of extortion known as "paying for protection."

Forced Choice

With forced choice, you select from options where the best choice is not immediately apparent. This method allows you to identify another person's trade-offs between conflicting issues by posing forced-choice questions. Analyzing these trade-offs can be advantageous in understanding preferences and decision-making processes. By studying trade-offs among a large group of people, you can inform your strategy and design a course of action.

One market research technique that utilizes this approach is conjoint analysis. Conjoint analysis helps refine the alignment of a new product with customer needs. It can also determine which products or services to add or remove from your product line. Additionally, it can help select the correct set of customer benefits, derive the optimal pricing for a product, and measure potential demand for a new product.

A biblical example of forced choice is the story of King Solomon (Kings 3:16-28). When two women claimed to be the mothers of a child, Solomon proposed cutting the child in half. He knew the true mother would rather yield the child to the other woman than see it harmed. This use of forced choice revealed the truth.

Forcing an Evasive Opponent to Stand and Fight

You have several methods available:

1. **Make Your Opponent's Life Difficult**: Create circumstances so challenging for your opponent that engaging in battle becomes the more appealing option. This strategy can include disrupting their supply lines, spreading disinformation, or leveraging psychological warfare to create stress and uncertainty.

2. **Surround and Cut-Off Retreat**: Encircle your opponent, leaving them no viable escape routes. Cutting off their retreat forces them into a position where they must confront you. This method can involve strategically positioning, mobilizing forces to critical locations, and blocking all possible exit points.

3. **Confront by Surprise**: Seize the initiative and catch your opponent off guard with an unexpected confrontation. This strategy can be a game-changer, disrupting their plans and forcing them to react hastily. Surprise can be incorporated in various ways, from sudden attacks to unanticipated maneuvers or exploiting moments when your opponent is unprepared or vulnerable. Each method aims to manipulate the situation, making direct engagement the more attractive or inevitable choice for your opponent.

Friendliness

Friendliness is a powerful tool for attracting others, especially when they perceive a benefit in reciprocating your warmth. For example, China, with its vast potential for economic growth, attracts numerous suitors whenever it displays signs of friendliness. This approach fosters relationships and opens opportunities for collaboration and mutual gain.

Frontal Attack

Richard H. Buskirk was quoted in a past edition of the Executive Strategies newsletter. The quote was from his book Frontal Attack, Divide and Conquer, the Fait Accompli & 118 Other Tactics Managers Must Know. Mr. Buskirk described how to stop a business opponent cold:

1. **Aim at Strength, Not Just Weakness**: Target your opponent's strengths to create a more significant impact.

2. **Request Research to Stall**: Use requests for additional research or information to delay proceedings.

3. **Give the Opponent a "Flat Tire"**: Disrupt their plans by packing the planning committee with incompetent individuals, demanding unrealistic deadlines, or using covert influence to ensure failure.

4. **Let the Situation Worsen**: Sometimes, it's strategic to allow the situation to deteriorate, waiting for a more opportune moment to

strike.

5. **Confront When Right and Politically Strong**: If you are correct and politically strong, confront the issue head-on. Keep this old adage in mind, "If you strike at a king, you must kill him."

6. **Sandbag**: Give your opponent the impression that you are less connected, knowledgeable, or intelligent than you are; then, be prepared to counterattack effectively.

7. **Demand It in Writing**: Insist on written documentation. Often, positions soften in writing, which can buy you additional time.

Frontal Torpedo

This strategy involves launching a direct and decisive assault on the core of your opponent's forces. It necessitates confronting your opponent head-on with the certainty of your success. This bold approach should be reserved for situations where you are almost guaranteed victory and can only be employed sparingly.

Gambit

A gambit is a strategic maneuver where one makes a calculated move to gain an advantage. A good example of this is the position that tobacco companies took on smoking before settling their major lawsuits. They encouraged smokers to dismiss the overwhelming body of medical research highlighting the numerous health hazards of smoking. Instead, they promoted smoking as socially glamorous, emphasized smokers' rights to smoke, and cited a few tobacco company-sponsored studies that contradicted some findings. In retrospect, these claims were merely a gambit to continue selling their cancer-causing, addictive products without restriction.

Gas Lighting

In this strategy, you make your opponent question their sanity, a tactic often called "gaslighting." To execute this, you align with allies—individuals your opponent trusts—to support your efforts in creating doubt and confusion in your opponent's mind. You can make your opponent feel disoriented and uncertain by consistently presenting false information and denying their reality.

For example, imagine a workplace scenario where a manager wants to discredit an employee. The manager enlists other colleagues to undermine the employee's confidence subtly. They might deny previously agreed-upon facts, question the employee's recollection of events, or subtly imply that the employee is overreacting or imagining things. Over time, the employee may start doubting their memory and perceptions, feeling isolated and vulnerable.

If successful, this psychological manipulation can cause your opponent to lose confidence and credibility, allowing you to achieve your goal without direct confrontation.

Get-Tough Policy

In this strategy, you leverage your resources and influence to intensify pressure on your opponent. This approach involves systematically applying more of your power to create an environment that increases their stress and difficulty.

For instance, in a fiercely competitive business scenario, you might ramp up marketing efforts, cut prices, or launch new products to outcompete a rival. Alternatively, in negotiations, you could introduce additional demands or deadlines to make the situation more challenging for your counterpart.

The key to this strategy is gradually escalating the pressure, forcing your opponent into a position where maintaining their stance becomes increasingly untenable. By continuously amplifying the intensity, you can compel them to yield or compromise, achieving your objective with a show of strength and determination.

Go-Between/Foil

In this strategy, you use a go-between to execute a portion of your plan, effectively disguising your overall strategy and protecting your objectives from being fully revealed. By not disclosing the entire objective to the go-between, you ensure that they:

1. Operate from a limited perspective

2. Cannot reveal information they do not possess

3. Achieve more within a restricted scope

This limited knowledge makes the go-between more likely to do their part without compromising the broader strategy. This method is particularly effective in negotiations, where you place someone between yourself and the opposing party. This intermediary can negotiate on your behalf without the authority to finalize agreements, leaving you the flexibility to:

1. Reject unsatisfactory terms.

2. Initiate additional negotiation rounds.

3. Choose or dismiss the final bidder based on your criteria, regardless of any compromises, promises, or agreements made by the negotiator.

For example, you might delegate the authority to negotiate but retain the power to commit or sign the agreement. By withholding final authority, you maintain control over the negotiation outcome. This approach also empowers your negotiator, who can make promises knowing that final approval rests with you.

Ultimately, this strategy allows you to guide the negotiation process from a distance, ensuring your overarching goals remain protected and adaptable.

Go For It / Just Do It

When information is lacking, rather than waiting passively for more details, take action based on your experience, intuition, and the available data. This approach carries higher risk, but the energy, flexibility, and speed you apply can mitigate other issues and make you more adaptable and open to change.

A good friend has long championed this philosophy, using the phrase "Just do it!"—well before it became synonymous with Nike®'s advertising slogan. She encourages anyone wavering or uncertain about starting a new endeavor to embrace this proactive mindset. It's excellent advice worth remembering.

Going for Specifics

A key goal in negotiation is to get your opponent to specify their detailed requirements first and as soon as possible. Negotiations are more manageable with specifics rather than generalizations. Push your opponent for details while responding with generalities.

The Cold War Russians often used this strategy. They publicly proposed arms reductions, creating a favorable impression, but consistently avoided dealing in specifics at the bargaining table. This approach was a powerful tool, allowing them to maintain a negotiating advantage by keeping their true intentions vague while pressing the other side for precise commitments.

Good Guy-Bad Guy

This strategy is commonly used in negotiations and interrogations. Two individuals decide beforehand which will play the role of the "good guy" and which will be the "bad guy." These roles are enacted once the negotiations or interrogations begin, without the opponent knowing that roles are being played.

The "good guy" is friendly and helpful during the discussions, while the "bad guy" is obnoxious, storms around the room, and may even walk out. The opposing forces create a tense atmosphere that can lead the opponent to:

1. Become confused or distracted

2. Experience stress, which they will try to alleviate, often by conceding

3. Align themselves with the "good guy" for relief

The "good guy" then leverages this bond to gain advantages in the negotiation. This technique manipulates the opponent's emotional state, making them more likely to cooperate or agree to terms favorable to the negotiators.

Gouge

You can exploit your opponent in various ways. In street fights of the past, individuals have been known to gouge out an opponent's eye. Car dealers often gouge prices on vehicles in short supply and high demand. During gas shortages, stations frequently gouge customers with inflated prices. Similarly, hoteliers have been known to gouge attendees of significant events, such as the Olympics, with exorbitant rates. Gouging occurs when you take unfair advantage or extract an unreasonably high price because you control a sought-after necessity. Some view this as intelligent business thinking under the banner of "supply and demand," but there is a point where such decisions cross the line from fairness to greed.

Gravamen

Gravamen is the critical juncture, pivot point, or crux of the matter. It refers to the crucial turning point when you shift from defense to offense. This turning point is both a specific location and a moment in time. The strategy involves deciding when to transition after being on the defensive.

Grudge

A grudge is a lasting malice resulting from an attack on your ego or a threat against you or your loved ones. It can sometimes arise as a delayed reaction. Holding a grudge means harboring an unresolved response to a threat or attack until the right time and opportunity for retaliation present themselves. Retaliation, the natural consequence of an attack or conflict, often stems from holding a grudge.

Gunboat Diplomacy

This strategy aims to influence someone's decision by demonstrating your power without using it. This approach relies on mental intimidation rather than physical coercion. For example, publicly threatening to fire the next employee who violates a key policy can deter potential violations and avoid the need for firings.

On a positive note, while interviewing a potential employee, you might have your secretary interrupt to sign several pay raises. This subtle display of rewarding performance can encourage the prospective employee to accept a lower salary, believing they can earn raises in the future.

Guilt

If you can create guilt through eloquent speech with a sympathetic person, you can later capitalize on that guilt with well-timed complaints. For example, if you remind your mother occasionally that she always favors your older sibling, you can leverage this when you want something from her. When the moment comes, you might say that if your older sibling asked for it, she would give in. She may consent if you don't overplay your hand. This ploy can be highly effective, allowing manipulators to use guilt repeatedly to get their way.

Gun to the Head

This strategy involves placing someone in a position where they must make a decision they are reluctant to make or to make it sooner than they want. For example, you suggest a friendly reorganization of a company while holding the majority of its common stock, effectively pressuring them into making a decision they would rather delay or avoid.

Hand-to-Hand Combat

This strategy involves engaging in direct, face-to-face combat, either one-on-one or against multiple opponents simultaneously. It is the essence of warfare and all offensive actions. Victory or defeat is determined when you or your resources confront and overcome, or are overwhelmed by, your opponent or their resources.

Use this strategy when:

1. **You Have a Clear Advantage**: Engage your opponent directly when you have superior strength, resources, or tactical positioning.

2. **Decisive Action is Needed**: When a quick and definitive resolution is necessary, confrontation can force a conclusive outcome.

3. **Surprise is on Your Side**: If you can catch your opponent off guard, direct combat can exploit their unpreparedness for a swift victory.

4. **All Other Options are Exhausted**: When negotiation, deception, or indirect methods have failed or are impractical, direct engagement becomes the last resort.

This strategy is high-risk but can be highly rewarding if executed correctly, as it focuses all efforts on a decisive point of conflict.

Hang One in the Yard

The basic premise is to execute an enemy and display their body where the rest of their allies can see. This action serves as a stark warning, deterring others from similar behavior and sending a clear message that whatever actions warranted such punishment should not be repeated.

Harassment

Harassment relies on a position of strength but is often temporary. A bully retains their power only until a stronger bully appears. On a larger scale, as a smaller competitor, your actions might harass a larger rival. If you start taking their customers and threatening their market position, they may retaliate and crush you. It's often wise to stay under the radar of more prominent competitors. While harassment can effectively weaken or distract an opponent, it also risks provoking a strong counterattack.

High Pressure Sales

When selling your plan, it's essential to be on the offensive. High-pressure salespeople use these techniques:

1. Touch people (Americans only), like tap an arm.
2. Avoid the color red.
3. Keep your message simple.
4. Think positively about your listener.
5. Nod subtly to affirm agreement.
6. Be aware of the effect of the weather on attitudes.
7. Keep your listener relaxed.
8. Get on the same level as your customer.
9. Make your listener feel special.
10. Position your client so they must look up at you.
11. Tell your listener a secret.
12. Appreciate your listener's strengths and blessings.
13. Be sensitive to your listener's mood.
14. Address every individual.
15. Preemptively address objections.
16. Maintain eye contact.
17. Say "I apologize" instead of "I'm sorry."
18. Emphasize your listener's importance.

19. Build a vision of a dream.

20. Encourage clear statements from your listener rather than simple responses.

21. Explain why you are the best.

22. Avoid arguments.

23. Be polite and listen.

24. Allow your listener some control.

25. Always be positive.

26. Use third-party stories to apply pressure.

27. Don't talk past the close.

28. Know your subject and be excited about it.

29. Know your listener.

30. Sell the benefits more than the product itself— sell the sizzle, not the steak.

31. Address all objections while moving toward the close.

High Volume Means Low Substance

An old Chinese proverb states, "Strong and bitter words signal a weak cause." Pay attention to the tone and volume of your opponent. Observe how vigorously they are advocating their position; this can provide insight into the strength of their argument. Conversely, when presenting your case, speak more softly than you think is necessary.

Identification

Identification involves linking your intended action to a specific objective. For example, declaring, "We're going to take that hill!" creates a commitment. Once the declaration is made, backing down or failing to succeed carries a penalty. If you enter a bank and announce you are robbing it, the consequences remain the same whether you change your mind or fail. Identification means tying your commitment to an action. This technique is used by extortionists and by anyone setting goals for themselves.

Incrementalize

As the saying goes, "get a toehold," "get a foot in the door," or "get a nose under the tent." The idea is to take a small piece of territory and then gradually expand your control. For example, an article in the November 2006 issue of Wired Magazine (page 48) described a technique for claiming the armrest on an airplane. Brian Lam, the writer, suggests reclining your seat a few inches. This tactic creates a small space behind the elbow of the person beside you. Place your elbow in that opening and gradually move the rest of your arm forward until you claim the entire armrest. This approach works on airplanes and in larger-scale scenarios, such as one country attempting to take over part of another.

Insanity Plea

Many people will excuse your behavior if you explain that you were incompetent or irrational when you misbehaved. Sometimes, individuals even feign insanity to avoid punishment.

Intensified Raid

An intensified raid is a small-scale attack executed when your opponent least expects it, designed to heighten their fears. To carry out such a raid effectively, you must thoroughly understand your opponent's entire layout and the status of their resources. Then, move in quietly and strike precisely where and when they are most vulnerable. Superior numbers on a micro-scale are crucial to minimizing the risk of failure in this type of operation.

In Their Face

This strategy involves boldly displaying your disregard for punishment in front of those imposing it. A prime example comes from U.S. professional football, where a coach attended a press conference wearing only a towel after being fined thirty thousand dollars for barring a female sports reporter from the locker room post-game. This action vividly illustrates the essence of the strategy.

Intentional Accident

This strategy involves indirect communication. For instance, you write a letter addressed to one vendor containing comments relevant to another. Then, you "accidentally" send this letter to the other vendor. This method obliquely delivers your message.

Intimidation

This strategy involves using your superior power to dictate the behavior of an opponent with clearly inferior power. It focuses on bullying, dominating, and exerting control over your opponent. However, this approach often generates resentment and can lead to retaliation when the opportunity arises.

Invasion / Direct Attack

This strategy involves a direct attack to penetrate your opponent's territory deeply. Initially, you may encounter strong resistance near the frontier, weaker resistance as you advance, and maximum resistance near the opponent's center of gravity.

Jack-in-the-Box

The Central Intelligence Agency employs this strategy, which involves using a dummy as a substitute for a rider in a car. The dummy acts as a decoy, allowing an agent to slip away unnoticed.

Kidnapping

Kidnapping involves physically taking someone against their will or seizing something and holding it for ransom. This narrow and illegal strategy is typically used to extort money from the kidnapped person's friends or relatives or to gain a political advantage. The FBI is highly skilled at capturing kidnappers in the United States, so this strategy should only be considered if you can defend the territory where the victim will be held against all threats and secure the ransom without being caught or traced.

In a broader sense, kidnapping can be applied in business by legally holding a project hostage rather than a person. In this context, it is a political maneuver. The counterstrategy is to treat this action like any threat and ignore it. Ignoring the threat is risky but can render the strategy ineffective. Your strongest position is genuinely not caring about the outcome. Your second strongest position is to feign indifference convincingly. If the kidnapper believes you do not care, their threat loses its power. However, they might still execute the threat if they think you are bluffing, which is a significant risk.

Another counterstrategy is to comply with the kidnapper's demands. This approach might help in capturing the kidnapper, but there is still a risk they will carry out their threat or, in some cases, return the hostage.

Tactics for Large vs. Small Companies

The strategies most commonly used by large companies against small companies are as follows:

1. **Predatory Pricing**: Large companies may lower their prices to unsustainable levels to drive new entrants out of the market. However, small companies can withstand this pressure by focusing on their unique value propositions and customer relationships. Once the small company withdraws or folds, the large company raises prices back to profitable levels. Although technically illegal, this is often defended as "defending our market." This tactic is common among large airlines in competitive markets like Hawaii.

2. **Volume Purchase Agreements**: Leveraging their purchasing power, large companies can negotiate lower costs than small companies. With these lower costs, they can price their products to drive smaller competitors out of business. While technically illegal, this practice is often defended by claiming that large orders create legitimate economies of scale, resulting in lower production costs. Consequently, volume discounts have become accepted practices that help keep prices down.

3. **Joint Ventures**: Large companies may form joint ventures with small companies, often with ulterior motives. During the

joint venture, the large company learns from the smaller one while developing a superior product independently. Once the new product is ready, the large company ends the joint venture, claiming the smaller company's product is no longer competitive. The smaller company's commitment to the joint venture often causes them to fall behind, leading to their exit from the market. The large company thus eliminates a competitor and shifts the blame onto the smaller company.

4. **Patent Enforcement**: Large companies often hold extensive patent portfolios and use them aggressively against smaller competitors. They may file lawsuits for patent infringement, knowing that small companies may lack the financial resources to fight lengthy legal battles. These lawsuits can drain the smaller company's resources, forcing them to settle, license the patents, or even cease operations. Large companies defend this tactic as protecting their intellectual property, but it can effectively eliminate smaller rivals.

5. **Exclusive Contracts**: Large companies may secure exclusive contracts with suppliers, distributors, or retailers, preventing small companies from accessing essential resources or markets. These agreements can limit the smaller company's ability to compete effectively.

Large companies often justify these contracts by highlighting the stability and guaranteed business they bring to partners, but the result is usually reduced market access for smaller competitors.

6. **Marketing and Advertising Dominance**: Large companies typically have substantial marketing and advertising budgets, allowing them to dominate media spaces and consumer mindshare. They can launch extensive campaigns that overshadow small companies, making it difficult for them to gain visibility. This tactic can be particularly effective in consumer markets where brand recognition plays a significant role in purchasing decisions, often leaving small companies struggling to compete.

7. **Regulatory Influence**: Large companies often have the resources to influence regulatory environments to their advantage. They can lobby for regulations that create barriers to entry or impose costs that small companies find challenging to bear. This influence can lead to a regulatory landscape that favors large incumbents and stifles competition from smaller firms.

8. **Mergers and Acquisitions**: Large companies frequently use mergers and acquisitions to eliminate competition. They may acquire smaller competitors to integrate their innovative products and

technologies or remove them from the market. This strategy consolidates market power and can discourage new entrants by demonstrating that the market is controlled by a few dominant players.

Tactics for Small vs Large Companies

Small companies often find themselves up against large competitors with significant resources. However, they can employ a variety of strategies to level the playing field and carve out a niche in the market. Here are some effective strategies for small companies to use against large companies:

1. **Differentiation and Niche Focus**: Small companies can thrive by offering unique products or services that cater to specific customer needs. By focusing on a niche market, they can provide specialized offerings that larger companies, with their broader focus, may overlook. This approach not only attracts a loyal customer base but also allows small companies to avoid direct competition with larger firms.

2. **Superior Customer Service**: Large companies often struggle to offer personalized customer service due to their size. Small companies can capitalize on this by providing exceptional, individualized service that builds strong customer relationships. This can include personalized communication, faster

response times, and tailored solutions, which can create a loyal customer base that values the personal touch.

3. **Agility and Innovation**: Small companies can be more agile and responsive to market changes than their larger counterparts. They can quickly pivot their strategies, adopt new technologies, and innovate faster. By staying ahead of market trends and continuously improving their offerings, small companies can maintain a competitive edge.

4. **Building Strong Local Ties**: Developing a strong presence in the local community can be a powerful strategy for small companies. Engaging with local events, supporting community initiatives, and building relationships with other local businesses can create a strong, supportive network. This local focus can help small companies gain loyal customers who prefer to support local enterprises.

5. **Leveraging Technology**: Small companies can leverage technology to streamline operations, reduce costs, and enhance customer experiences. By adopting digital tools and platforms, they can compete with larger companies on efficiency and innovation. This can include using customer relationship management (CRM) systems, e-commerce platforms, and digital marketing strategies to reach and engage customers effectively.

6. **Strategic Partnerships and Alliances**: Forming strategic partnerships with other small businesses or complementary companies can provide mutual benefits and enhance competitiveness. These alliances can offer access to new markets, shared resources, and collaborative innovation. By working together, small companies can create a stronger collective presence in the market.

7. **Emphasizing Sustainability and Ethical Practices**: Many consumers today prefer companies that prioritize sustainability and ethical practices. Small companies can differentiate themselves by adopting environmentally friendly practices, supporting fair trade, and promoting social responsibility. This can attract customers who value these principles and are willing to support businesses that align with their values.

Late Strategy

Senator Frank Church employed this strategy when he announced his belated candidacy in the 1976 U.S. presidential campaign. His plan was to exploit voters' indecision, as none of the official candidates were clearly succeeding. By entering the race late, he aimed to capitalize on voter fatigue with the existing candidates. He hoped that a string of victories in the late primaries would position him as the most credible candidate at the convention.

However, since he did not come close to winning, there is speculation that his effort might have been a ploy by his party. The party, it is suggested, may have used Church as a pawn to attract voters who were disillusioned with the leading candidate from the opposing party, thereby weakening the opposition and drawing some of their supporters into Church's camp.

Lead-In/Entrapment

This strategy resembles the Drawn-in/Trap strategy (see above) but relies on genuine attraction rather than trickery or diversion. The lead-in serves as an attention-grabber, designed to draw an opponent closer. An effective attention-getter focuses attention on the desired outcome. For instance, you might use a photograph of an elaborate, expensive service to highlight the value and soundness of your more affordable service proposal. In some cases, your opponent can be led into a trap by pursuing the attraction. This strategy requires careful prior planning on how you will trap your opponent.

Leak / Premature Certainty

This strategy involves leaking a premature position statement that appears definitive while being reasonably confident about the expected outcome. By doing so, you can gauge the reaction. If the response is negative, you can quietly abandon the issue.

Leak **/ Trial Balloon**

This strategy is similar to the previous one but involves even less certainty. With a trial balloon, you leak an idea to gauge reactions without knowing how people will respond. This approach is suitable for:

1. Assessing sensitivity to an unknown issue.

2. Determining or beginning to build acceptance for an idea or proposal.

3. Deflecting criticism to others.

Letting Them Negotiate Against Themselves

Instead of making a counterproposal, suggest they go home to "think about it and return with a better offer."

Leveling Uneven Odds

You can level the playing field when outnumbered by employing tactics such as mobility, sneak attacks, and superior weaponry. Utilizing mobility allows you to maneuver quickly and unpredictably, making it harder for the opponent to pin you down. Sneak attacks take advantage of surprise, catching the opponent off guard and disrupting their plans. Superior weapons can give you a technological or strategic edge, enabling you to inflict greater damage with fewer resources. By combining these strategies, you can effectively counterbalance the numerical advantage of your opponent.

Leverage

This strategy aims to create a sense of obligation in your opponent. Salespeople often employ this tactic by taking potential clients to lunch or a sporting event, fostering a feeling of obligation and, consequently, loyalty. Small, thoughtful gifts can also be used to obligate someone. The key to success is ensuring these gifts are personal (but not intimate), as this personal touch can significantly enhance the recipient's sense of obligation and connection.

Logic

You should employ this strategy against logical opponents if you are confident that your logic is superior or you can catch them off guard with an illogical move when they expect a logical one. This approach can exploit their reliance on predictability and create opportunities for unexpected advantages.

Logical Approach

This strategy targets opponents who think logically. By understanding their logical approach, you can predict their behavior and use this predictability to your advantage. By anticipating their moves, you can devise tactics that exploit their expectations and create opportunities to outmaneuver them.

Loss of Face

To maintain balance, always provide your opponent a way to save face, even in total defeat. Failing to do so can lead to deep-seated feelings of revenge, potentially culminating in future retaliation. The greater the loss of resources and dignity, the stronger the opponent's desire for intense vengeance. Conversely, treating your defeated opponent with respect and honor can lead to their more willing acceptance of your control and reduce the likelihood of future conflicts.

Lynching

Lynching refers to enacting your form of justice outside the bounds of existing law, typically involving taking someone out and hanging them. It often occurs when a mob believes that regular justice will not sufficiently punish the perceived crime. Participants in a lynching party risk being charged with a crime themselves. In a business context, the term "lynching" can apply when mob sentiment is directed toward a single person or program. It is possible to orchestrate such a mob action remotely by appealing to weak minds and inciting them to act on your behalf.

Magnifying a Problem

This political strategy, often employed by the press, involves identifying a problem with an opponent and excessively covering it to create a significant news story. For instance, when one U.S. government administration left the White House offices slightly trashed as a "dirty trick" on the incoming administration, the press attempted to escalate the event into a federal crime. Similar tactics are seen in every election, where the press seeks to magnify minor flaws in candidates' characters into significant news stories. This approach highlights the difference between reporting genuine news and manufacturing news from trivial details.

Manhunt/Search

An organized search involves dedicating significant resources to hunting down and capturing, removing, or destroying a single individual. This strategy extends beyond pursuing an escaped prisoner; it can also encompass an all-out effort to target a single politician or any person obstructing your business's advancement.

Maneuver Attack

This form of attack involves luring your opponent into a vulnerable position for a strike. The maneuver positions your opponent where you want them and exploits their mistakes to your advantage. This strategy typically includes the following tactics:

1. Targeting the opponent's resource stores.

2. Preventing the unification of two opposing forces.

3. Serving as a threat during retreat.

4. Attacking isolated, poorly defended territories.

Manipulating a Person's Thinking

Manipulation encompasses tactics such as half-truths, lies, evasiveness, and emotional distancing. On a mass scale, manipulating public perception is essential for effective propaganda. However, here, we focus on manipulation on an individual level. Manipulative strategies often involve flattery and feigned interest or pursuit. If you recognize someone is trying to manipulate you, the best countermeasure is to break off the relationship. Alternatively, understanding and directly addressing the underlying motive can also neutralize the manipulation. Generally, ending an unhealthy alliance is the most effective course of action.

To counter a manipulative person, consider the following strategies:

1. Share information about yourself to relax them.

2. Ask them about themselves to build rapport.

3. Identify common interests.

4. Treat everyone around them equally to avoid favoritism.

5. Show genuine liking for them, as people tend to reciprocate positive feelings.

6. Remove the pressure of decision-making from them.

7. Develop a friendship that goes beyond mere cordiality.

8. Mirror their movements and gestures to create a sense of connection.

9. Deliver a sharp, clear, and engaging presentation to keep them alert.

10. Use various emotions to build a sense of closeness and encourage impulsiveness.

Manipulation without Alienation

Here are some practical methods for persuading people to do what you want:

1. **Inspire sympathy**: "Our [such and such] will [suffer] if you don't want to [do as we ask], and we're already in enough trouble."

2. **Ask for a return favor**: " Remember when [we did this for you]? Well, now we need [something] from you."

3. **Imply prior agreement**: " In your work on project X, [you agreed to . . .]."

4. **Request an exception to the rule**: "As you know, we normally wouldn't [ask for this], but it is imperative that [we have it now so that . . .]."

5. **Offer support**: "This is a top priority. Can we get someone else to help you? It's crucial for [the success of the project]." " Compliment

occasionally: "How about applying those great skills of yours to [help us out]?"

6. **Appeal to mutual benefit**: " By [helping us with this], you'll also gain [specific benefit]."

7. **Create a sense of urgency**: "We need [this] completed by [time/date] to avoid [negative consequence]."

8. **Highlight a common goal**: "We both want [desired outcome], and [doing this] will help us achieve it."

9. **Offer incentives**: "If you [do this for us], we can provide [incentive] in return."

10. **Invoke authority or credibility**: "Experts in [field] agree that [doing this] is the best approach."

11. **Present it as a challenge**: "I know you're capable of [achieving this], and it would be a great accomplishment."

12. **Use peer pressure**: "Everyone else on the team has agreed to [do this], and your support would be valuable."

13. **Show appreciation in advance**: "Thank you so much for considering [helping us with this]. Your assistance means a lot to us."

14. **Frame it as an opportunity**: "This is a unique chance to [achieve/experience something], and your involvement would be crucial."

Mediating Conflict

Richard Darman, who served as Deputy Secretary of the Treasury from 1985 to 1987 under President Reagan, said, "True innovations of politics are transacted when the final, crucial words of an issue are put down on paper." In an interview with Francis X. Clines, Darman outlined several key elements of mediation:

1. **Compromise**: Both sides must be willing to make concessions.

2. **Power Diffusion**:Ensure that power is distributed and balanced.

3. **Share**:Distribute both credit and blame appropriately.

4. **Time**: The more time invested, the quicker progress can be made.

5. **Trial Balloons**: Test ideas to gauge reactions, determine positions and isolate issues through trial and error.

6. **Conflict Resolution**: Resolve conflicts by addressing and correcting wrongs.

7. **Move to Action**: Pragmatically translate philosophy into action to advance the process.

True politics requires managing diverse constituencies, including bosses, subordinates, customers, suppliers, unions, and peers. The challenge is to achieve cooperation and ensure smooth progress simultaneously.

Mediator—When to Use or Not

Experts recommend using a mediator under the following circumstances:

1. **Continued Relationship**: Both parties wish to maintain their working relationship. Lawsuits often damage relationships beyond repair.

2. **International Disputes**: The parties are in different countries, and international lawsuits can be prohibitively expensive. Conversely, mediation is a cost-effective solution that can save both parties significant financial resources.

3. **Large Projects**: The dispute involves a significant project where delays could harm the project's progress. Lawsuits would only exacerbate these delays.

4. **Short Product Life**: The product or process has a short lifespan, and a lawsuit would likely take longer than the process itself. Mediation, with its focus on swift resolution, is a more efficient option in such cases. For instance, General Motors (GM) has employed independent arbitration to resolve customer complaints. Their program is designed to be free, quick (resolving issues in 40-47 days), and does not require lawyers. It is binding only on GM, aiming to reduce the number of lawsuits.

However, mediation might not be suitable in certain situations, such as:

1. **Lack of Willingness**: When one or both parties are unwilling to participate in good faith.

2. **Power Imbalance**: When there is a significant power imbalance between the parties, it could result in unfair outcomes.

3. **Legal Precedents**: When establishing a legal precedent is essential.

4. **Severe Misconduct**: In cases involving severe misconduct or criminal activity where legal intervention is necessary.

Mess with their Heads

If you can execute an action against your opponent that they foresee but are powerless to prevent, you will significantly boost your team's morale while demoralizing your opponent's. This tactic demonstrates your strategic superiority and ability to control the situation, reinforcing confidence within your ranks and instilling a sense of inevitability and defeat in your adversary.

Murder / The Kill

You might be surprised to learn that killing can be considered a strategy. It serves both as a strategy and a counterstrategy. When premeditated, cold-blooded, or without a clear motive, it functions as a strategic move. Conversely, it becomes a counterstrategy when used for retaliation or revenge. As a strategy, killing is employed when one player sees a clear path to victory, and the consequences are either irrelevant or non-existent, as often seen in wartime scenarios.

Killing can also be a post-strategy, a last resort when all else fails. For example, someone who suffers a total loss might resort to murdering the perceived victor. This extreme measure underscores the desperation and finality of their situation.

Mutual Damage

Labor strikes, price wars, and boycotts are all examples of mutual damage. In a labor strike, workers forgo their wages while the company loses the services of its employees, causing financial harm to both parties. During a price war, a company lowers its prices to gain market share, only to be undercut by competitors. This tactic often leads to a downward spiral where both sides incur significant financial losses. Eventually, the company with the deepest pockets typically prevails and raises prices even higher than before. Similarly, a boycott involves consumers depriving themselves of a product while depriving the company of sales.

Most forms of mutual damage are a form of political pressure aimed at forcing someone to change their position. These actions often arise from philosophical disagreements and carry the potential for significant change, albeit with harm to all involved parties.

Neutralize

If you cut off your opponent's power source, you effectively end their ability to oppose you. Neutralizing critical resources as you achieve intermediate objectives further weakens your opponent's resistance capacity. Systematically dismantling their support structures reduces their overall power and control.

Neutral Out

"To neutral out" involves escaping an impossible situation by engaging a neutral third party to arbitrate a binding resolution. This approach is beneficial in conflicts where the involved parties cannot agree. The third party, acting as an impartial mediator, helps to facilitate a fair and balanced solution that both sides must accept. This method not only resolves the immediate issue but also helps to maintain relationships by providing an unbiased perspective. Engaging a neutral arbitrator can bring about a legally binding resolution that ensures a just outcome, fostering trust and cooperation among the parties involved.

Newspaper Attack

You can leverage a newspaper article or public announcement with a strategic lead-in that disguises your bias while attacking your opponent. For example: "It's not the first time the people of [specific country] have been oppressed by [opponent]." Or, "Once again, [country] has engaged in [specific action] to intimidate [regional group] striving for their rights to [objective] and equitable trade with [country]." Such statements assume the reader already agrees with your perspective. They project a tone of righteousness and strength, sounding both positive and critically appropriate, even when overtly biased. This technique uses the veneer of objectivity to influence public opinion and subtly reinforce your viewpoint.

Occam's Razor

Occam's razor is a principle attributed to William of Ockham, a 14th-century English philosopher and theologian. It states that the simplest explanation is usually the correct one. In strategic contexts, this means using the simplest effective strategy until proven inadequate.

William of Ockham (c. 1287–1347) was a medieval scholar known for his contributions to logic, theology, and philosophy. He advocated for simplicity in explaining phenomena, arguing that unnecessary complexity should be avoided. This principle, now known as Occam's razor, is foundational in various fields, including science, philosophy, and strategic planning.

In practice, applying Occam's razor involves starting with the most straightforward approach that can achieve the desired outcome. If and when this approach fails, more complex strategies should be considered. This approach helps maintain efficiency and clarity in decision-making processes.

"Opinionators"—Dealing with Outside Opinions

If a consultant or auditor comes to review your operation, here are some recommendations sourced mainly from the Research Institute of America:

1. **Cooperate Fully**: Embrace oversight from top management and actively participate, demonstrating your willingness to cooperate.

2. **Evaluate Credentials**: Assess the reviewer's qualifications and their capability to judge your operation. Ensure they are competent.

3. **Collaborate Closely**: Work alongside the reviewer to foster a cooperative environment and ensure accurate assessments.

4. **Clarify Scope**: Understand the reviewer's scope and assignment and ensure they stay within these boundaries.

5. **Be Transparent**: Answer their questions honestly without volunteering excessive information. Address real problems without overwhelming them with issues you are already solving.

6. **Seek Assistance**: Guide the reviewer towards areas where you could benefit from top management's greater exposure and support.

7. **Adapt to Their Style**: Understand the reviewer's approach and style. If they prefer analysis, provide them with relevant material to analyze.

8. **Avoid Blame**: Do not blame others for uncovered issues. Instead, commend the reviewer for identifying problems and addressing them immediately.

9. **Implement Recommendations Promptly**: Accept and begin implementation without waiting for the final report. If recommendations are unreasonable, discuss them directly with the reviewer.

10. **Document Your Context**: Provide context to ensure issues are viewed correctly, considering mitigating circumstances like budget cuts or staff reductions.

11. **Seek Support**: Enlist help from other well-positioned individuals to strengthen your position.

12. **Stay Professional and Positive**: Maintain a professional demeanor and positive attitude throughout the review process to foster a constructive relationship with the reviewer.

13. **Follow-Up**: After the review, follow up on the recommendations and communicate with the reviewer or audit team to demonstrate your commitment to continuous improvement.

14. **Prepare in Advance**: Before the review, ensure that all documentation and processes are up-to-date and ready for examination. Preparation can significantly influence the outcome of the review.

Addressing these points ensures a thorough, cooperative, and effective response to a consultant or auditor review. The only caveat is to respect your own experience and knowledge and be prepared to challenge any recommendations you disagree with.

Opponent Reaction

Every action has a consequence, and your opponent's reaction is that consequence in competitive situations. An opponent's response can vary simultaneously in two dimensions: the intensity of the reaction and the timing. The intensity can range from total submission to violent retaliation, while the timing can be immediate or delayed. Immediate submission may lead to violent reprisals later when the opportunity arises. Every action elicits an opposite and equal or greater reaction, and the only way to avoid this reaction is to take no action.

Optics

Optics is a strategy that creates the illusion of reality, presenting something as more favorable than it is. For example, a store offering furniture with no money down and no interest for twelve months uses optics to attract customers. However, the underlying terms may involve deferred interest or other conditions. In business, a form of debt instrument might be used to inflate the purchase price of a takeover, with the expectation that the debt will be paid from the acquired company's assets. These strategies give the impression of advantageous terms or financial stability while concealing underlying complexities and risks.

Organizational Development

Every strategy requires organization. Relevant organizing strategies include:

1. **Management by Objectives**: Clearly define individual goals that align with the overall objective. This approach effectively communicates direction and is most successful with genuine delegation.

2. **Team-building**: Use team-building as a motivational tool to direct attitudes and bring out submerged opinions, fostering a cohesive and collaborative environment.

3. **Norm Modification**: Identify weaknesses within the organization and either convert them into strengths or eliminate them.

4. **Job Enrichment**: Organize tasks so that individuals can complete entire tasks rather than following an assembly-line approach, increasing job satisfaction and productivity.

5. **Socio-Technical Redesign**: Provide feedback on progress or accomplishments to alleviate the monotony of repetitive jobs, enhancing job satisfaction and efficiency.

6. **Situation Analysis**: Understand the critical factors and their relationships and how to resolve, defuse, and eliminate issues to improve decision-making and strategy execution.

These strategies ensure that goals are clear, teams are motivated, weaknesses are addressed, tasks are enriching, feedback is constructive, and situational factors are thoroughly analyzed.

Outnumbering the Opposition

This strategy is employed when you bring more of your supporters to a meeting than your opposition. By doing so, you can influence the discussion so that the majority of opinions, comments, and support are in favor of your objectives.

Package Trade

The Package Trade involves consolidating several desired elements into a single opportunity, leveraging synergism—where the combined value exceeds the sum of individual parts. For example, you might offer three separate bidding opportunities to a single vendor in exchange for lower prices. While each account alone might not justify discounts, the combined volume and the prestige of securing a larger sale make it worthwhile for the vendor to offer reduced prices. This approach creates a win-win situation, maximizing value through strategically bundling opportunities.

Paintbrush

This strategy is highly devious. It starts with a broad, critical accusation. When the accused challenges this broad criticism, you exploit the opportunity to present specific circumstances that are not entirely related to your initial comment. By providing evidence supporting these particular instances, you create the impression that all your comments are true. Even if the original broad accusation was false, this method leaves a lingering impression of its truth. This strategy is particularly effective for individuals who do not think critically enough to recognize the manipulation.

Persistence

Even if your proposal is initially rejected, wait a while, repackage it, and try again. Persistence is respected, especially when your proposal has some truth or merit. This approach demonstrates your commitment and willingness to adapt, increasing the chances of eventual acceptance.

Persuasion

To make your efforts to persuade more effective, consider the following steps:

1. **Keep Your Presentation Simple and Well-Structured**: Clear and concise presentations are easier to follow and more persuasive.

2. **Stick to the Facts**: Present factual information and let the audience draw their

conclusions.

3. **Summarize and Ask Clearly for What You Want**: Conclude with a summary of the key points and clearly state your desired outcome.

4. **Anticipate Objections**: Conduct dry runs with high-level peers to refine your approach for potential objections.

5. **Include Sacrificial Elements**: Build-in components that can be conceded without undermining your core proposal.

6. **Be Willing to Concede Points:** Understand that you may need to compromise on some points to achieve overall success.

7. **Meet Resistance with Flexibility and Cooperation**: Respond to pushback with a willingness to adapt and collaborate.

8. **Maintain Proposal Integrity**: Avoid conceding so much that your proposal becomes unviable. Instead, withdraw and develop it further if necessary.

9. **Prioritize the Company's Interests**: Consider the company's broader interests to demonstrate alignment with organizational goals.

10. **Maintain a Positive Attitude**: Approach everyone involved positively and respectfully to build goodwill.

Additional relevant steps include engaging the audience with questions, anecdotes, or examples to keep them invested in your presentation and using visual aids to reinforce key points and make complex information more accessible. Evidence such as data, case studies, or testimonials helps build credibility. After the presentation, follow up with the audience to address any lingering questions or concerns and reinforce your key points. Knowing your audience and tailoring your presentation to their interests and concerns can significantly increase your impact.

Persuasion succeeds based on the completeness of your presentation and how well it addresses the "what's in it for me" motive of the person you are attempting to persuade. Groups can be more challenging to convince due to differing participant goals, but appealing to a group's shared motives or goals can make them more receptive. Be aware that unknown factors may prevent acceptance of your proposal, so it is wise to informally discuss your proposal with key constituents beforehand to gauge their reactions and refine your approach.

Physical Pain

The application of physical pain to extract information from unwilling opponents is often associated with torture. It is generally applied in an escalating pattern, making the victim anticipate unbearable pain, thereby pressuring them to reveal concealed information. Physical pain can also serve as punishment in some instances. However, as the Geneva Convention prohibits this approach, it is highly controversial and not recommended for solving problems. Torture is illegal, unethical, and may not work, but some argue it could be justified in extremely rare and severe military situations.

Using physical pain as a means of extracting information or as punishment is not only inhumane but also often ineffective. It can lead to false confessions or unreliable information, as individuals under extreme duress may say anything to stop the pain. Moreover, the use of torture can have severe psychological effects on both the victim and the perpetrator, leading to long-term consequences for individuals and societies.

Given its legal and moral implications, alternative methods of interrogation that respect human rights and dignity are strongly advocated and more effective. Building rapport, psychological tactics, and strategic questioning are preferred techniques that align with ethical standards and international laws.

Pincer Movement / Pincher Action / Pinch

A pincer action is a strategy where two forces converge and concentrate their power on a single point of attack or opposing force. This approach involves attacking the opponent from at least two directions that meet at a central point. By applying pressure from multiple sides, the opponent is surrounded and constrained, making it difficult for them to maneuver or defend effectively.

This concept can be applied in various ways in business. For example, you can leverage the pressure of a hired firm with a fiduciary responsibility to ensure that a deal with a third party is fair and transparent. This process is mandated in England for large construction projects, where an assurance firm is hired to verify that the contractor is not cheating the client.

Overall, the pincer action is a versatile strategy that, when applied effectively, can lead to significant advantages by overwhelming the opponent or ensuring fair practices in business transactions.

Placing in Play

This strategy involves making an action, such as a takeover attempt, public knowledge. The objective is to attract additional participants to the action. This approach can also serve as a defensive measure when seeking a "white knight"—an ally who can save you from a hostile takeover or other threats. Publicizing the situation increases visibility and creates opportunities for potential allies to step in and offer assistance, thereby strengthening your position and potentially deterring adversaries.

Playing or Stopping Company Politics

Playing politics is different from being politically astute. Playing politics involves management engaging in noncooperative behaviors as they compete for higher organizational positions, often with a selfish orientation. This practice involves a variety of underhanded tactics aimed at gaining personal advantage, often at the expense of others and the organization's overall goals. Such strategies focus on noncooperation and self-interest, significantly harming organizational culture and effectiveness.

Strategies for playing politics include:

1. **Backstabbing**: Undermining colleagues to advance one's position.

2. **Maneuvering**: Strategically positioning oneself to gain an advantage.

3. **Manipulating**: Influencing situations or people for personal gain.

4. **Doing an End Run**: Bypassing proper channels to achieve a goal.

5. **Infighting**: Engaging in internal conflicts within the organization.

6. **Shifting the Blame**: Redirecting responsibility for mistakes onto others.

7. **Covering Your Behind**: Taking actions to protect oneself from blame or criticism.

8. **Building a Case**: Collecting evidence to support or discredit a position or person.

9. **Railroading a Decision**: Forcing a decision through without proper consideration.

10. **Fighting the Organization**: Opposing organizational policies or decisions for personal gain.

11. **Defending Your Turf**: Protecting one's area of influence or responsibility.

12.**Building a Power Base**: Establishing a support network to increase personal power.

13.**Stealing Credit**: Taking credit for others' work or ideas.

14.**Defending Self-Interests**: Prioritizing personal interests over organizational goals.

15.**Turf Protection**: Safeguarding one's area of control from encroachment.

16.**Spying/Espionage**: Gathering confidential information about colleagues or competitors.

17.**Burying Someone's Career**: Deliberately undermining a colleague's career progression.

18.**Keeping "Book" on Someone**: Documenting a colleague's mistakes to use against them later.

19.**Spreading Rumors**: Disseminating false or misleading information to damage someone's reputation.

20.**Withholding Information**: Deliberately not sharing critical information to gain an advantage or undermine others.

21.**Forming Alliances**: Creating informal groups to consolidate power and influence decisions.

22.**Sabotaging Projects**: Deliberately undermining projects led by others to make them look incompetent.

23.**Exploiting Weaknesses**: Identifying and taking advantage of the weaknesses of others to advance one's position.

24. **Feigning Support**: Pretending to sincerely support initiatives or colleagues while secretly working against them.

25. **Leveraging Favors**: Using past favors to manipulate others into supporting your agenda.

26. **Playing the Victim**: Exaggerating personal hardships to gain sympathy and support from others.

27. **Gaslighting**: Manipulating someone into doubting their perceptions and judgments.

28. **Cherry-Picking Data**: Selectively presenting information supporting your narrative while ignoring data contradicting it.

29. **Gatekeeping Resources**: Controlling access to essential resources to maintain power over others.

30. **Engaging in Flattery**: Using excessive praise to gain favor and influence decision-makers.

Each of these strategies focuses on noncooperation and advancing one's interests at the expense of others. This behavior can overshadow customer needs, ignore competitor actions, and significantly reduce organizational effectiveness. Playing politics diverts focus from performance, cooperation, and teamwork, causing rumors to become more important than actual intelligence. Personal goals often conflict with organizational goals, draining the organization's energy and hindering its progress.

If you are competing against an opponent with severe internal political issues, you can gain ground quickly. Conversely, if your organization suffers from internal politics, it is crucial to address the problem swiftly. This can be achieved by refocusing on performance, competitors, and organizational goals. Reorganizing and placing better people in key positions can also help mitigate the negative effects of internal politics.

Strategies to mitigate the impact of internal politics include fostering transparency by encouraging open communication and reducing the spread of rumors and misinformation. Promoting collaboration creates a culture of teamwork and cooperation, counteracting individualistic behaviors. Setting clear policies on ethical behavior and conflict of interest, encouraging accountability by holding individuals responsible for their actions, and providing leadership training to help leaders recognize and address political behaviors within their teams are all effective measures.

By effectively addressing internal politics, an organization can maintain a focus on performance, cooperation, and achieving collective goals. By doing so, you can steer the organization back on track and ensure its long-term success.

Political Astuteness

Political astuteness involves being sensitive to the unique personalities and communication styles of those above you in the organization and being attuned to their directions and goals. This skill emphasizes a teamwork orientation, focusing on cooperation and alignment with organizational objectives.

The top leader is responsible for curbing political games within the organization. Practical strategies to stop politicking include:

1. **Focus on Cooperation and Teamwork**: Prioritize collaboration and long-term goals over individual competition.

2. **Share Credit and Recognize Accomplishments**: Ensure that contributions are acknowledged and celebrated.

3. **Reward Based on Results**: Provide timely and clear rewards for achieving results.

4. **Give Clear Direction**: Assign roles and responsibilities with clarity to avoid ambiguity.

5. **Communicate the Leader's Vision**: Ensure everyone understands the leader's expectations and vision.

6. **Encourage Change**: Promote regular change and adaptability within the organization.

7. **Offer Opportunities for Growth**: Provide new challenges and growth opportunities for employees.

8. **Rotate Jobs and Encourage Education**: Implement job rotation, training, and

educational opportunities to broaden skill sets.

9. **Search for the Right People**: Continuously seek and hire individuals who align with the organization's values and goals.

10. **Encourage Innovation and Confidence**: Foster a culture of innovation, inspire confidence, and involve everyone in the organization's mission.

11. **Provide Clear Feedback**: Ensure employees know where they stand through regular feedback.

12. **Communicate Effectively**: Engage in two-way communication, emphasizing listening more than talking.

13. **Maintain Core Business Philosophy**: Be flexible in approaches but keep the fundamental business philosophy intact.

14. **Show Empathy**: Care about employees, their families, and their challenges.

15. **Explain the "Why"**: Clarify the reasoning behind directives to enhance understanding and buy-in.

16. **Promote Informal Social Activities**: Encourage fun and social interaction in and out of the workplace.

17. **Understand Personal Needs**: Be aware of and respond to employees' personal needs at all levels.

18. **Stay Informed**: Stay connected to informal

communication channels, like the "coconut wireless" or rumor mill.

By implementing these strategies, leaders can foster a positive, cooperative, and politically astute organizational culture prioritizing performance, teamwork, and mutual respect.

Plausible Deniability

This strategy protects high-ranking leaders from being held accountable in delicate situations. It involves withholding vital information from the leader, allowing them to later truthfully claim they were unaware of the matter. Leaders cannot be held responsible for something they did not know about.

Underlings, who learn the sensitive information and choose to keep it from the leader to protect them, usually do so to provide a level of protection. If the withheld information becomes widely known and problematic, one of these loyal subordinates typically volunteers to take the blame for not informing the leader. The volunteer is often discreetly rewarded, maintaining the respect and privacy in the leadership dynamics.

This approach ensures that the leader remains untainted by controversies, maintaining credibility and authority while the organization navigates the issue.

Play on Ego

To get what you want or to attain the support

of a self-absorbed person, uses compliments, deference to their authority, intelligence or wisdom and kowtow to his or her inflated self-image.

Play on their Vulnerabilities

A technique used to get captives to talk, playing to their vulnerabilities means gaining their trust or using their ego to get them to reveal what they fear or feel the most guilt about and then use that information to extract more information from them.

Positioning

In 1561, Ruy Lopez, a Spanish priest and one of the earliest recorded chess players, theorists, and writers, ascribed a strategy where you "place your opponent with the sun in his eyes." In other words, try to set up the playing field to your advantage. Another form of positioning is to mentally prepare a person in a higher authority position that a proposal will soon be presented to them. By doing so, you get them in the proper frame of mind and even give them a chance to prepare. Another common way to assert dominance over someone, like a job applicant, set your chair higher than theirs.

Power Firing

If you have the power to fire people from their positions, then you can use this power to send a signal. You can send one kind of signal to an entire group by firing an incompetent. You can send another kind of message by firing a dissident. An opponent's failure to fire is a weakness of which you can take advantage. Low performers are usually well known. Use this fact to point out the weakness of your opponent's decision-making. In firing someone, do not be apologetic but rather candid, unemotional, and a bit cool.

Preemptive Decision

The objective of this approach is to keep decisions to yourself. This result is achieved by not permitting decisions to be made by those below you. You reprimand them severely whenever they make a decision, which should have considered other factors or other departments. Thereby, through fear and pressure, you control the flow of information and set yourself up as the "Grand Decider." By so doing, you establish control and exercise your power. However, by so doing, you also assume full responsibility for any wrong decisions.

Promises

Promises are a fundamental strategic tool. They can buy time and come in various forms: false promises, half-truths, and genuine commitments. Promises can influence finances, reputation, or status, implying gains or losses. A promise of gain appeals to the recipient, while a promise of loss acts as a threat. An example of a threat promise is a penalty clause in a contract. Promises can be used to save face or to embarrass, to foretell promotions or demotions. They can be integral to strategies that position someone as an opponent who may not yet recognize it.

Protecting Your Flank

Flank protection is crucial when "proceeding in depth." This term refers to penetrating deeply into your opponent's territory with a column of forces. Without adequate flank protection, any part of this column that gets severed could be jeopardized, compromising the entire operation.

Provocation

This strategy involves a deliberate show of force. By intentionally provoking a dispute and forcing a showdown, you demonstrate your toughness and control, but only when you are confident of victory.

Pseudo Center on Gravity Strike

Attacking your opponent's center of gravity compels them to concentrate their forces for increased protection. By deploying a small unit to strike at this critical point, you can then:

- Attack other objectives with your larger forces.

- Target the opponent's resources as they move toward the center.

- Follow the small attack force with a large-scale assault.

- Exploit the distraction to your advantage.

Pumping for Information

Pressuring people for information is a strategy to obtain data that might otherwise be inaccessible. Classic police movies often depict detectives badgering suspects under glaring lights. Techniques range from spies like Mata Hari using seduction to thugs employing pure extortion. The key is to apply extreme pressure to force the opponent to reveal the needed information. This approach can include using reverse psychology, exploiting a weak self-image, or issuing threats. It's important to note that some forms of this strategy are illegal and carry serious consequences, such as threatening someone or their family.

Purple Hearts

Purple Hearts are awarded for wounds received in battle, recognizing those who have suffered losses in pursuing the group's objectives. By awarding Purple Hearts, you acknowledge and honor the sacrifices made. This strategy provides recognition, builds motivation, and increases loyalty within the group.

Putting the Fox in Charge of the Henhouse

This technique involves placing the person who threatens the organization in a position of responsibility. This action is sometimes done out of ignorance, but it can also be intentional, as in "giving them enough rope to hang themselves."

Pyramidal Pricing

This pricing strategy involves setting prices based on the customer's ability to pay. It stratifies customers according to their financial resources, forming a pyramid structure. At the top are the fewest customers with the most ability to pay, while at the bottom are the most numerous customers with the least ability to pay. IBM has successfully employed this strategy for its software for years.

Psych-Out

This strategy leverages actions and words to communicate on a subconscious emotional level with your opponent, suggesting their inadequacy. For example, you might convey the message, "others think you are a nobody," to make your opponent preoccupied with others' opinions, undermining their self-confidence and distracting them from their plan. The goal is to cause mistakes and distractions by psyching out your opponent. The strategies include:

1. **Provocation**

   - **Cold Shoulder (Shunning)**: This involves deliberately ignoring or excluding the opponent. It's particularly effective on novices or naive individuals who may become demoralized or frustrated when they feel overlooked or unimportant.

- **Teasing**: Light-heartedly mocking or goading the opponent can push them to try harder than necessary, leading to mistakes. This method leverages the opponent's pride and competitive spirit against them.

- **Cage Rattling**: This tactic involves friendly kidding or taunting about one's abilities to unsettle the opponent. For example, asking, "Did you correct that slice in your golf swing?" can introduce self-doubt and disrupt one's focus.

- **Teacher-Teacher**: Condescendingly offering subtle advice can undermine opponents' confidence, making them second-guess their strategies or techniques.

2. **Intimidation**

- **Legal, Physical Violence**: This involves using permissible aggressive tactics to intimidate the opponent, such as tackling the quarterback extra hard in football to make them wary and hesitant.

- **Secret Weapon**: Revealing a new, unexpected tool or strategy, like a high-tech golf ball, can intimidate the opponent by making them feel unprepared and at a disadvantage.

3. **Evoking Guilt Feelings**

- **Mr. Nice Guy**: Acting overly friendly and nice can lull the opponent into a false

sense of security, causing them to relax
and lower their guard, even when winning.

- **Poor Soul**: Highlighting your handicaps
  or disadvantages can evoke sympathy from
  the opponent, making them less aggressive
  or competitive.

- **I Do Not Care**: Pretending to play for fun
  rather than competition can deflate the
  opponent's sense of victory, diminishing
  the value of their win and making them
  feel guilty for taking the game too
  seriously.

4. Distraction Derby

- **"Look, Over There"**: Diverting attention
  away from the main issue by pointing to
  something else, thus shifting focus.

- **Get Angry at Someone, Anyone**:
  Using temper tantrums or anger to create
  chaos and distract from the real issue.

- **Gee Whizzing**: Overloading the
  opponent with trivial or irrelevant details
  to confuse and distract them from the
  main issue.

- **Information Overload**: Bombarding
  the opponent with excessive data and
  details, making it difficult for them to
  focus on what's truly important and
  effectively counter your actions.

Quick Strike

The Quick Strike strategy involves executing a rapid and decisive attack to overwhelm your opponent. This strategy is most effective against an opponent with superior resources who is not yet organized or one with inferior resources. In either case, the key advantage lies in the speed of the strike—catching your opponent unprepared for the suddenness and force of your attack. This surprise element disrupts their plans and leverages the element of shock to gain the upper hand.

Quid Pro Quo (Something for Something)

You've likely heard the saying, "You scratch my back, and I'll scratch yours." This principle forms the basis of bartering. Trading equal values depends on the specific wants or needs of the traders involved. Such exchanges are crucial in politics, business, and personal relationships. Always ensure you receive something in return for what you do for others, immediately or in the future. Likewise, remember that when someone does you a favor, they will likely come back to collect on it someday.

Quinine

Quinine is a strategy that is unpleasant (i.e., a bitter taste) but necessary and yields beneficial results. For example, this could involve having to fire a poor performer who also happens to be your friend.

The Rabbit

In competitive running races, the term "rabbit" refers to a non-competitive runner strategically used to set a fast pace for the other competitors. This tactic is commonly employed in middle-distance and long-distance races to ensure a fast overall race time and to help competitors achieve personal bests or record-breaking performances. Here's how the "rabbit" tactic works:

1. **Selection of the Rabbit**: The rabbit is usually a skilled runner capable of maintaining a predetermined pace for a portion of the race. This runner is typically not competing for the win and is often compensated for their role.

2. **Setting the Pace**: At the start of the race, the rabbit takes the lead and runs at a fast, consistent pace. This lead helps the competitive runners stay on target for their desired finishing times. The rabbit's role is crucial in the early stages of the race, where maintaining an even pace can be challenging due to adrenaline and the dynamics of the starting pack.

3. **Psychological Benefits**: By following the rabbit, competitive runners can focus on their form and conserve mental energy, as they don't have to worry about pacing themselves. This benefit allows them to stay relaxed and maintain a steady effort, knowing that they are on track for a fast time.

4. **Drafting Advantage**: Running directly

behind the rabbit provides a drafting benefit, reducing air resistance and making it easier for the following runners to maintain the pace. This technique can be particularly advantageous in windy conditions or in longer races where conserving energy is vital.

5. **Rabbit's Exit**: The rabbit usually exits the race after fulfilling their pacing duties, typically around the halfway mark or slightly beyond, depending on the race distance and the agreed strategy. The competitive runners then take over, ideally having been set up for a strong finish with a consistent, fast pace already established.

6. **Strategic Implementation**: The rabbit tactic is often used in races where record attempts are made, or competitors aim for personal best times. It is less common in tactical races where the outcome depends more on strategic positioning and finishing speed than overall time.

7. **Organizational Support**: Race officials or individual athletes' coaches often organize the use of rabbits, ensuring that the race conditions are optimized for fast performances. The rabbit's role is well-communicated and integrated into the race strategy.

Rallying Point

Exploiting an emotional event or landmark can be an effective way to garner a following. For instance, the rallying cry "Remember the Maine" used an emotional situation to stir people to action. Similarly, Hindus in India are using the claim of an ancient temple taken from them 450 years ago to rally support against Muslims.

Rapid Apology

If you are under someone else's control or have made a mistake, and they know you are wrong, apologize quickly and sincerely. Generally, they will respond with understanding. However, if you deny your mistake despite their awareness, you will lose credibility and respect in their eyes.

Reconnaissance

A small force is deployed to gather intelligence on your opponent or execute a minor attack, compelling them to reveal their strengths. This strategy aims to determine your opponent's size, location, strength, or hidden motives.

Red Herring

The term "red herring" originates from the practice of using smoked fish to train hounds by distracting them with the scent, thus diverting them from the intended trail. In strategy, a red herring involves using misleading bait to divert your opponent, causing them to waste valuable resources and time by pursuing a false lead.

Referendum Manipulation

In political strategy, parties often place emotionally charged referendums on the ballot to maximize voter turnout. These referendums are carefully selected to ignite strong feelings among the party's base, motivating them to participate in the electoral process. The underlying strategy is that while these voters cast their votes on the passionate issue, they are also likely to support their party's candidates in other races. This approach leverages single issues to drive broader electoral engagement, ensuring a higher turnout for the party's candidates.

For example, a referendum was introduced in Florida to mandate more room for pregnant sows. The need for this measure stemmed from growing concerns about animal welfare and the conditions in which farm animals were kept. Animal rights activists highlighted the cramped and inhumane conditions in which pregnant sows were confined, often in gestation crates so small that the animals could not turn around. This issue resonated strongly with liberal voters passionate about animal rights and welfare.

By placing this emotionally charged referendum on the ballot, the Democratic Party aimed to galvanize these voters, increasing turnout. The strategy proved effective, as the referendum passed and brought many liberal voters to the polls, almost resulting in a victory for Al Gore in the state.

This example illustrates how single-issue referendums can be used to drive broader political engagement and support for party candidates.

Release Valves

Offering choices can be beneficial when making tough proposals. Providing options allows strong objectors to avoid taking a rigid negative stance. By presenting alternatives, you enable them to express preferences and find common ground, which can lead to more constructive discussions and potential compromises. This approach reduces resistance and makes decision-making smoother, as people are more likely to engage constructively in negotiations when they feel their input is valued and considered.

Relentlessness

Relentlessness involves returning repeatedly until your opponent loses the will to resist. The strength of your will is the decisive factor.

Repetition

Repetition is a powerful tool for persuasion. Presenting the same message or proposal multiple times can convince people of various arguments over time. In some cases, only a few repetitions are needed to be effective.

Reprisal

When you hold a position of dominance and strength, it is crucial to maintain that maximum strength. Any behavior that threatens your position must be addressed swiftly and decisively. Reprisals, or returning injury for injury, are necessary for actions that oppose your authority or purpose.

Reputation Impairment and Entrapment

By employing embarrassing or illegal tactics, you can place your opponent in a compromising position, thereby discrediting them. These methods expose their vulnerabilities and tarnish their reputation, making it difficult for them to maintain credibility and authority. The resulting scandal or legal trouble can significantly weaken their standing and diminish their influence.

Restructuring

A straightforward way to neutralize a threat from someone at a lower level in your organization is to initiate a restructuring that eliminates their position, causing them to leave.

Retaliation

Retaliation occurs in direct proportion to the degree of suppression that causes it. The greater the suppression, the stronger the retaliation. Additionally, the more intense the suppression, the longer a person will hold onto the need for a resolving response—retaliation. Suppression creates an emotional imbalance that can only be resolved through retaliation or appeasement. Sometimes, the threat of retaliation is enough to elicit appeasement, provided the threat is credible and substantial.

Retreat

Retreat, a strategy in itself, is typically a defensive strategy aimed at conserving and refocusing resources. When executed effectively, it can serve as an offensive tactic if the reason for retreat is to reposition or reorganize for another attack. This effectiveness of retreat can preserve strength and prepare for future opportunities, instilling confidence in the team.

Reverse Apology

Use an apology to highlight a position. For example, "I'm sorry that my religion has become an issue in this election but . . ." Or, for another example, during a 1984 presidential debate, Ronald Reagan diffused concerns about his age by humorously stating, "I will not make age an issue of this campaign. I am not going to exploit, for political purposes, my opponent's youth and inexperience."

Road Net

This strategy focuses on routes of influence—understanding who influences whom. Additionally, it encompasses the network of roads, determining the number, speed, and direction from which replenishing resources will arrive.

Rock Soup

Rock soup, a strategy involving a step-by-step escalation of activities, is a concept that might surprise you. In warfare, it could begin with initial reconnaissance, followed by reinforcement, and culminate in a full-scale attack. This strategy gradually builds up pressure, keeping the opponent casual and unprepared for as long as possible. The term' rock soup' originates from a story told by General George S. Patton about a tramp who asked for water to make rock soup. After receiving the water, he added two polished white rocks and then requested potatoes and carrots for flavor, eventually acquiring meat as well.

The moral of this story highlights two fundamental elements of effective strategy:

1. **Gradual Approach**: Escalate slowly to minimize emotional reactions.

2. **Disguised Intentions**: Conceal true intentions to keep defenses down.

Russell's Law

The greatest wars often have the smallest beginnings. This principle, known as Russell's Law, emphasizes the importance of understanding the potential consequences of any action you take, as it may escalate into something much larger. Named after the British philosopher Bertrand Russell, this concept highlights how minor events can lead to significant, unforeseen outcomes. Always consider the broader implications of your actions to prevent unintended escalation.

Saber Rattling

This strategy employs intimidation by showcasing your strength to threaten your opponent. For example, during the Cold War, the Russians held May Day parades to demonstrate their world-class weapons of mass destruction to the United States and the world.

Sabotage

This strategy involves covertly deploying trained resources behind enemy lines to damage and destroy your opponent's assets. The objective is to weaken the opponent or delay their advance by cutting off supply lines, destroying supplies, and incapacitating transportation and vehicles.

Salami Cuts

Adlai Stevenson first used this strategy to reduce the Federal budget by making small cuts in various areas. The approach involves cutting funds here and there, but ultimately, no one is appeased or satisfied. Consequently, this strategy rarely succeeds.

Salvage Sale

The best time to buy is when the second person goes broke, attempting what the first person already failed at, as this leads most people to believe that success is impossible. A second failure creates a highly pessimistic mindset, presenting a unique opportunity.

Sandbag

This strategy involves holding back strength when your opponent believes you have exhausted your resources. By doing so, you can surprise and potentially overcome them. This tactic acts as a form of decoy, allowing you to hold back a position now to strike later with multiple strategies, greater strength, or at a better opportunity. The key here is control. If you can control the timing and if additional time does not benefit your opponent, only then should you defer action until later. This control empowers you to make strategic decisions that can turn the tide in your favor.

The counterstrategy is to force the issue—make your opponent act now by attacking, cornering, or driving them to where they must respond.

A notable example is Andres Gomez's explanation of reaching the 1990 French Open finals. He described his win over Thomas Muster, who had beaten him earlier at the Italian Open: "In Rome, I did not throw him everything I had. It is not that I did not want to win, but I wanted to save a few shots against him here in case I had to play him." —a classic use of sandbagging.

Another example is: During the 1972 World Chess Championship, Bobby Fischer employed sandbagging against Boris Spassky. Fischer deliberately lost the first game and didn't show up for the second game, forfeiting it. This unusual behavior baffled everyone, including Spassky. Fischer's odd tactics put Spassky off balance, but Fischer came back strong, winning the match 12.5 to 8.5. Later, Fischer joked, "I like the way they look at me. It's like I'm going to kill them or something."

Scapegoating

In this strategy, you select a politically weak individual from your group and set them up to "take the rap." With enough support from others, blaming an innocent, naïve, or helpless person for a mistake or crime becomes relatively easy. A typical example is firing a subordinate to shield higher-paid elites at the top of an organization from blame or culpability. The targeted person is sometimes chosen out of malice as payback for a prior offense or affront.

There are several counter-strategies: act quickly to uncover and disclose the motives of the perpetrators, confront them to make clear that you know what they are doing, talk openly about it until it stops or concludes, and avoid being defensive by confidently sticking to the facts. If the inevitable happens, accept it with grace.

Second Position

Sometimes, starting in the second position can provide a significant advantage. The resort development business often demonstrates this strategy: by waiting for the first developer to go broke, you can step in and take over what they started. Similarly, runners frequently nominate a "rabbit" in a foot race to set the pace and lead for most of the race. This tactic lets you watch the leader blaze the trail and clear obstacles, giving you a strategic advantage.

Secrets of Selling Big Ideas

In nearly every case, a strategy needs to be sold to someone. Here are some ideas from the Research Institute of America:

1. **Start with a Small Pilot Project**: To bolster assumptions and refine the plan, begin with a pilot.

2. **Present without Distractions**: Choose a time when other distractions are absent.

3. **Align with Listeners' Interests**: If known, match the proposal to the listeners' interests. Creating interest in something new is challenging unless you build on existing interests.

4. **Do Your Homework**: Research every assumption thoroughly, study similar undertakings in-depth and understand the payoffs and timelines. Test the sensitivities of assumptions and gauge the listener's

reactions.

5. **Provide Supporting Success Stories**: Show how other leaders are pursuing the same idea and obtain endorsements from reputable people.

6. **Involve the Listener**: Engage the listener and pull them into the project.

7. **Observe and Listen Carefully**: Listen carefully to reactions, and let others explore the negative aspects before responding. Avoid responding defensively.

8. **Present in Stages**: Study, test, and phase each new idea to keep risk low.

9. **Use Clear Visuals and Data**: Enhance your presentation with clear visuals and data to illustrate key points and make the information more accessible.

10. **Be Prepared for Questions**: Anticipate potential questions and have well-thought-out answers ready to address concerns and objections.

By following these steps, you can effectively sell your strategy and gain the necessary support.

Selective Targeting

When engaging a force, prioritize targeting those aiming directly at you; those shooting wildly can wait. Differentiate your enemy's resources and focus on those most critical to your opponent or dangerous to your assets.

On a larger scale, when facing a substantial force, attack their leaders first. Removing their leadership disrupts the central command and can demoralize and disorganize the larger force, potentially causing it to lose its will to fight.

Self-Aggravating Action

This strategy involves taking action with no consequence, which creates a situation of significant consequence. For example, a rumor of food shortages can lead to a rush on grocery stores, ultimately causing a real shortage.

Set Up for a Later Fall

In this strategy, you employ a subtle or less-than-obvious action designed to maneuver your opponent into a more vulnerable position. This action may create chaos, havoc, or confusion, or it may be intended to lead your opponent into a trap. Alternatively, it could give your resources the "taste of a win."

Shadow Candidate

The Shadow Candidate tactic involves running an independent or same-party candidate with a similar name to a major candidate in an election. The aim is to siphon off votes from the target candidate by creating confusion among voters. This tactic can be particularly effective in closely contested elections where even a small number of diverted votes can impact the outcome.

Shills

This strategy involves using a collaborating partner to entice another into action. The old-time auctioneer's trick of employing a friend to bid up prices is a classic example of a shill. Las Vegas casinos use this tactic by adding their people to poker games as shills (they will identify them if asked). Similarly, a bartender's tip jar, primed with a $1 bill, serves as an inanimate version of a shill.

Shotgun

The shotgun strategy involves multiple attacks, timed divergently to confuse and overwhelm the opponent. Typically, it starts with a decoy attack, followed by second and third attacks of equal strength from different angles and paths. For example, an initial feint might distract the enemy's forces in a military operation, followed by simultaneous strikes on their supply lines and command centers, creating chaos and reducing their ability to mount a coordinated defense.

Siege

A siege is a prolonged operation, usually a gradual attack that reduces the risk of disaster. It may involve a minor conquest to collect tradable trophies or a sustained effort to wear down the opponent over time. In business, a siege might occur when your execution is superior, but you do not overtly capitalize on it. For example, you have well-integrated customer databases and order processing systems, resulting in faster and more reliable order fulfillment. However, you don't promote this advantage, gradually attracting new customers while keeping the reason hidden from your competitors.

A siege is a long-term strategy. Your advantage accumulates slowly as you wear down the enemy. However, you remain vulnerable because your opponent may receive reinforcements.

Some perceive a siege as an attack by an aggressor on a fixed defense, such as a fort. This fixed defense could be at your opponent's territorial boundary or within their territory. A siege may also entail surrounding your opponents and cutting off their supply of life-sustaining resources (i.e., food, water, electricity, medical supplies, healthcare, air, fuel, communications, shelter, clothing, sanitation, security, and safety).

Silent Guns

Originating from game theory, the term "silent guns" describes a situation where decisions must be made with incomplete information. This scenario occurs when you need to decide whether to shoot at your enemy without knowing if they have already fired at you. It's akin to a duel where the guns make no sound or flash. The strategy, developed by John von Neumann and Oskar Morgenstern, involves proceeding a certain distance toward your opponent, prepared to shoot at random intervals. This approach maximizes your chances of success despite the uncertainty.

Skirmish

A skirmish is a sudden, hostile, and brief encounter between opposing forces. It is a small-scale battle where the outcome is unlikely to devastate the combatants.

Skirmish Line

The skirmish line is a method used to protect officers on the front line. It involves sending "clouds" or "waves" of troops ahead of the main force, creating a buffer between your main force and the enemy. This tactic helps shield the primary contingent from direct enemy engagement.

Smear Campaign

Smearing an opponent involves producing and distributing unfavorable publicity about them. This strategy is primarily used against public figures; with private individuals, care must be taken to avoid slander. A smear is safest when based on embarrassing facts. However, it can also be effective with blatantly false information if presented by someone who appears trustworthy. When using false information for a smear, targeting your opponent's greatest strength can cause the most damage. For example, a politician with a distinguished military career can be attacked by a seemingly credible person who falsely accuses them of dishonorable conduct. Even without detailed evidence, such a charge can put the opponent on the defensive.

The purpose of a smear is to assassinate a

person's character or, at the very least, create
suspicion and doubt about them.

Smoke Out

This strategy involves driving your opponent
out into the open or forcing them to reveal their
location or strategy using external means. The goal is
to irritate, distract, or otherwise provoke them to act
against their better judgment or deviate from their
intended game plan.

Smoke Screen

In World War II, ships often laid down smoke
to cover the advance of personnel landing carriers.
Similarly, you can employ this strategy by disguising
your actions with a distraction that interferes with
your opponent's vision or diverts their focus from
their primary target. A diversion can keep attention
fixed in one direction while you act in another,
effectively concealing your true intentions.

Spoiling Attack

A spoiling attack is designed to create
apprehension, eliminate an objective's appeal, or
serve as a diversion. This tactic may involve targeting
a defenseless or neutral group to plunder resources
or capture essential supplies. Such attacks seem
illogical and can unexpectedly broaden the conflict.
Often, they are surprise attacks on unsuspecting
opponents, striking at nerve centers thought to be

impervious, dramatically impacting morale.

Sports-Washing

Sportswashing involves sponsoring or hosting major sporting events in your country to divert public and international attention away from other activities or issues that may be controversial or negative. By associating your country with high-profile sports events, you can enhance your image, promote positive narratives, and overshadow any undesirable news or situations.

Spread Eagle / Straddle

This strategy involves overlapping lease expiration dates when a single tenant leases multiple commercial offices, ensuring continuous occupancy. Another version involves creating contracts based on achieving extremes and securing payoffs when results are high or low but not in the middle. Conversely, a related strategy involves getting a payoff for middle results but not for extremes. For example, in a game between the Philadelphia Eagles and the Chicago Bears, you could arrange bets to win regardless of the outcome by leveraging extreme fan loyalties.

Spying

One of the oldest strategies involves sending operatives behind enemy lines to gather information. Operatives must speak the language fluently, have a solid backstory, convincingly carry out the deception, access critical areas, and have an escape plan.

Steamroller

The steamroller strategy involves decisively choosing a course of action and relentlessly pursuing it, letting nothing stand in your way. However, avoiding pursuing a path so doggedly is crucial that you neglect to reassess your actions.

Standoff

A standoff occurs when actions, strategies, or counterstrategies fail to help either opponent advance, resulting in a deadlock.

Stick by Your Guns / Stick to the Plan

As Admiral David Farragut said during the Battle of Mobile Bay, "Damn the torpedoes, full steam ahead." This strategy involves adhering to the planned course when events deviate from expectations. Using all available intelligence helps determine whether to stick to the original plan or deviate. Success requires information, experience, and boldness. Often, it's better to avoid overreacting and stick with the original plan long enough to get back on track, assuming the plan was well-conceived.

Straw Man

One version of the straw man strategy is to put forth a "trial balloon," a preliminary proposal to gauge reactions. If accepted, you can proceed further. Another version involves using a dummy or decoy to draw attacks, causing your opponent to expend resources while conserving yours. However, if the opponent discovers the trick, they may launch a full-scale assault in retaliation, which may or may not align with your expectations.

Strike at Their Fears

Understanding and exploiting your opponents' greatest fears can be a powerful strategy. During World War II, General Patton required his troops to keep their bayonets affixed because he knew the enemy feared them. Knowing what your opponent fears allows you to leverage that fear effectively.

Submissive Gestures

Submissive gestures involve displaying intentional weakness to force action. This strategy can place you in a dependency relationship, obligating your opponent to provide for your needs against their will. Alternatively, a submissive gesture can lure an opponent close for a surprise attack.

Suicide Mission

A suicide mission involves an attack on the center of gravity by resources with total abandon, expecting to die in the attack. Sacrificing one's life for a cause is the primary reason for this strategy, but such sacrifices are rarely worth the cost.

Surprise: As a Tool of Persuasion

In demanding negotiation situations, four types of surprises can be remarkably effective. However, their impact is greatly enhanced when they are delivered with a warm and friendly attitude. The four types of surprises are:

1. **Sudden Concession**: Unexpectedly giving up a bargaining chip.

2. **Political Shift**: Throwing your support behind someone you previously opposed.

3. **Change in Subject**: Altering the agenda to disrupt the flow and redirect the conversation.

4. **Mood Change**: Shifting from formal to informal, becoming agitated after being calm, or changing the atmosphere (e.g., cracking a joke, taking off your jacket, sending out for food, etc.).

These surprises can often lead to new concessions from the other party.

Surrender

The participant employs this strategy in the worst position when further resistance is futile. It becomes viable when both players recognize that the best alternative is ending the "game" without annihilating the other. In rare instances, the game can conclude with simultaneously eliminating both opponents. For example, there have been boxing matches where both fighters knocked each other out simultaneously.

Survival in Election Politics

The critical elements in election politics include the following:

1. **Public**: Generally passive and not interested in abstract problems, preferring simple issues. Most are concerned with self-interest and local matters such as taxes and crime.

2. **Process**: Understand the game's rules and the history of what has and hasn't worked. Stick to generalities and avoid specifics. Do not apologize for derogatory labels; instead, accept and redefine them to your advantage (e.g., "Sure, I'm a liberal if that means someone who cares about old people.").

3. **Packagers**: Every campaign needs "handlers" to conduct extensive research, manage exposure, and plan responses. Focus on what people are against more than what they are for.

4. **Polls**: Polls can indeed take on a life of their own. Leading in the polls can create momentum, as some people may swing your way. But do people choose their candidate based on what polls say? Some may, perhaps

to feel aligned with the majority. However, many voters do not base their decisions solely on poll results. Polls can be helpful to gauge campaign progress and public opinion trends, but they do not directly influence most voters. Voter influence typically comes more from a candidate's image, communication skills, and, to a lesser extent, their stance on issues.

5. **Press**: The press often focuses on daily snippets of candidates hurling abuses at each other. Detailed discussions are rare, with debate answers rarely exceeding 120 seconds. While some news agencies try to report substantive issues, most voters ignore the details.

6. **Perception**: Voter perception of a candidate's personality, integrity, and relatability plays a crucial role. A candidate's ability to connect emotionally with voters often outweighs policy specifics.

7. **Promotions**: Campaign advertisements and promotional materials must be consistent and resonate with the public's emotions and core concerns.

8. **Performance**: Public speaking and debate performance are critical. Candidates must present themselves confidently and convincingly to gain voter trust and support.

Tactics for New Small Businesses

When trying to break into a market, consider

the following strategies:

1. **Offer More Elegant or Elaborate Packaging**: Make your product stand out visually to attract attention and convey quality.

2. **Avoid TV Commercials**: TV advertising can be expensive and may not provide the best return on investment for new market entrants.

3. **Avoid Research and Development** (unless well-defined and attainable): Focus on leveraging existing technologies or innovations unless you have a clear and feasible R&D plan.

4. **Avoid Volume-Based Dealer Incentives**: Larger competitors can easily outbid you. Instead, offer graduated dealer incentives based on percentage increases in purchases, such as "Increase purchases by 10 percent, get this incentive."

5. **Be a Parasite**: Live off the crumbs of the big sharks by targeting niche markets and underserved segments that larger competitors overlook.

6. **Compete and Grow through Better Service and Lower Costs**: Focus on providing superior service and operational efficiency rather than merely lowering prices.

7. **Leverage Digital Marketing**: Utilize social media, SEO, and online advertising to reach your target audience cost-effectively.

8. **Build Strong Relationships with Early Adopters**: Identify and cultivate relationships with early adopters and influencers who can help promote your product.

9. **Focus on Customer Feedback**: Actively seek and incorporate customer feedback to improve your offerings and build loyalty.

10. **Create a Strong Brand Identity**: Develop a compelling brand story and identity that resonates with your target market.

Taking the Blows

They may release their pent-up anger by allowing an aggressor to "blow your socks off" a few times. This tactic can change their attitude, making them more receptive to your ideas and willing to listen better after venting. A similar technique is found in martial arts, where a martial artist stays close to their opponent instead of moving back completely when struck. By standing their ground and absorbing the blow, the martial artist can strike back more effectively and potentially demoralize the opponent. This strategy serves both offensive and defensive purposes.

Terrorism

Terrorism requires a purpose, a target, a location, and a method. In 1988, the primary method of worldwide terrorism against the United States was bombing (70 percent). Bombing is often perceived as a cowardly approach, as it is safe for the perpetrator, akin to shooting someone in the back. Other methods included armed attacks (12 percent), arson (8 percent), and kidnapping (3 percent). Each method has its benefits and drawbacks, including the risk of punishment when caught. Since 1988, terrorists have shifted to more extraordinary self-sacrifice in their attacks, with the cowardice now falling on commanders who ask others to die for their beliefs while not doing so themselves. Primary terrorist targets include businesses, government officials, diplomats, and military personnel, with prime geographic locations being Latin America, Asia, the Middle East, and Western Europe. Typical targets within these areas include aircraft, offices, cars/buses, people on the streets, and gathering places. The purposes of terrorism include creating fear among the general population, retaliation for perceived injustices, power struggles, religious furtherance, government destabilization, regional insurgency, fundraising, and seeking independent homelands. While terrorism may achieve short-term goals, it fails to create lasting change. Progress inevitably comes from dialogue and negotiation, with terrorists usually being caught and jailed or killed.

Testing a Lie / Silent Treatment

This strategy works best when trying to determine if a person is lying. By holding back and not responding to the suspected lie, you will find that the person may continue to talk, often only when they are lying. They will try to bolster their lie by adding more details. They will have nothing more to say if they are telling the truth.

Test Challenge

This strategy is used to test the mettle of your opponent. It involves issuing a challenge, dare, or act of defiance. Your opponent's response will indicate submission or defiance unless you withdraw the challenge. Your challenge may also catalyze an unorganized coalition to join together in a concerted defense.

The Ambush and the Location to Strike

Deciding "where to strike" involves choosing between confrontation and a stealthier approach. This choice fundamentally alters your plan. Engaging an opponent directly demands maximum strength, while approaching from behind emphasizes cunning and stealth. Each tactic requires different strengths and strategies.

The Ambush and the Time to Strike

Timing is a delicate balance between two extremes of preparedness. You can be underprepared, and you can overprepare to the point of diminishing returns. The optimal time to strike may even arise before you achieve full preparedness. Striking at the precise moment is challenging, but it's a balance that depends on several factors:

1. **Difference in Preparedness**: Assess the relative preparedness of your forces compared

to your opponent's.

2. **Opponent's Vulnerability**: Identify and exploit weaknesses in the opponent's positions.

3. **Resource Fluctuations**: Consider the potential for timely increases or decreases in the opponent's resources.

4. **Psychological Conditions**: These play a significant role in determining the optimal timing for action. Exploiting psychological factors, such as fears and morale, can be a game-changer. For example, in World War II, General Patton recognized two key fears within the German army: fear of the dark and fear of bayonets.

Understanding and balancing these factors can help you determine the optimal timing for action.

The Big Showdown

The big showdown is a direct, frontal assault where you risk everything for one final victory. Although rare, when such a decision arises, it is monumental. Total victory or defeat is uncommon, but individual competitors like boxers, wrestlers, and Tae Kwon Do practitioners understand its significance. Life-and-death competitions, such as war, duels, or Wild West gunfights, represent the ultimate showdown. Typically, both victory and defeat are fleeting. If you lose today but survive, you can fight again another day. If you win, you become the target for future challenges.

Defeat rarely happens as one big event. Instead, it often occurs as a slow succession of failures or losses. Critical decisions or failures to recognize the conditions leading to defeat usually precede the actual failure by an extended period. Entire industries, such as steel, textiles, and electronics, have collapsed over time due to a lack of recognition and response to adverse conditions. The big showdown is more honest—you know your situation and can confront your enemy immediately and directly.

The False Negative

The false negative strategy involves sending a misleading message to make your opponent misinterpret your intentions. A typical example of this tactic is seen in politics: the false negative campaign. In such a campaign, you disclose a fact that provokes your opponent to attack you on that point. When the attack occurs, you are not only prepared to respond and turn the point to your advantage but also use the opportunity to raise other issues you wish to discuss.

The Impossible Weakness

The most impossible place to attack is often the least defended, making it the weakest spot in the defense. This paradox arises because defenders typically allocate their resources and fortifications to areas they perceive as likely targets. As a result, less attention is given to less vulnerable areas, inadvertently creating weak points. Recognizing and exploiting these weak spots can lead to significant advantages, as surprise and minimal resistance can turn the tide in an otherwise challenging confrontation. This principle underscores the importance of a comprehensive defense strategy that considers potential vulnerabilities, even in seemingly secure areas.

The Setup

One of the plots from a MAS*H TV show features B.J. Honeycutt betting everyone that he can prank each of them by morning. Confident that he won't succeed, each person disputes his claim. One by one, however, B.J. successfully tricks them. As Hawkeye Pierce watches his friends fall victim to B.J.'s pranks, he becomes increasingly vigilant and paranoid. Finally, Hawkeye is the only one left to be pranked. Determined to avoid being caught, he stays awake all night.

At breakfast, Hawkeye, exhausted and bleary-eyed, proudly declares that B.J. failed to prank him. As Hawkeye is convinced that he has won the bet, everyone reveals they were in on the scheme all along. Hawkeye was the only one not in on the joke. He had stayed up all night for nothing; the ultimate prank was on him.

This example of a setup is perfect with these critical components:

- A premise or challenge

- Several players are in on the plan, and one subject is unaware

- The execution of the plan

- Revelation: The unaware player discovers they have been duped

- This strategy is fundamental to some confidence games

The brilliance of this setup lies in the unsuspecting participant's increasing paranoia and eventual realization that they were the target all along, demonstrating the power of collective deception and psychological manipulation.

The Larger, the Fewer

Strategies targeting the most significant rewards are rare. Big moves, often made by the highest-level individuals, require substantial resources, which are limited. Few potential participants are involved in starting major wars or high-stakes games because such conflicts demand significant opponents and extensive resources.

Threats

A threat is a communication of one's intentions emphasizing the automatic consequences of the other person's actions. It is a declaration that you will act, not that you may act, in a specified manner depending solely on your opponent's behavior. Ideally, a threat contains no suggestion that you might not follow through. Any hint of uncertainty allows your opponent to gauge the likelihood of you carrying out the threatened action.

The key to a successful threat is maximum credibility. You must minimize any room for judgment or discretion in carrying out the threat. The purpose of the threat is to relinquish your choice and eliminate alternatives to control the other party's behavior. Threats are akin to promises in that they both involve a commitment, but with negative rather than positive consequences. Additionally, a threat is costly if it fails, while a promise is costly if it succeeds.

In the context of threats, you are making a promise where the outcome depends on the other person's actions. The second party determines the result, and as the first party, you are committed the moment you issue the threat or commit to action.

Threat— Reactions/Counterstrategies

A threat strategy may elicit various reactions or counterstrategies, including:

- **Submission**: leading to buried resentment that may surface later

- **Negotiation**: seeking a middle ground

- **Avoidance**: escaping the situation entirely

- **Violent Retaliation**: immediate or delayed

The response depends on how the opponent assesses the strength of your position and the amount of time they have to respond. Additionally, threats are highly influenced by the recipient's mental state. The strength of the threat also affects the degree of response. A refreshed person is more likely to retaliate, while an exhausted person is more likely to submit, especially if the consequences of not submitting are clear and immediate.

Threat—First-Mover Advantage

One advantage of using a threat is the ability to move first and control the flow of action. By moving first, you can often gain the upper hand. Issuing a quick threat can limit your opponent's initial move to a few constrained choices. With some preparation, you can constrict, restrain, or even prevent your opponent's first move by doing the following:

- Anticipate the most likely first move and its consequences.

- Quickly devise a threat that diminishes your opponent's incentive for choosing that move, ideally defining the remaining set of choices for your opponent within the threat.

- Act to deter, distract, or disrupt your opponent's plans.

Your opponent may anticipate your first move based on your rules, habits, traditions, and other factors. If they can predict your behavior, moving first allows you to cut off their initial move and reduce their options. However, if your strategy relies on secrecy and surprise, moving first might not be advantageous.

Threats—The Size Effect

The size of a threat significantly influences your opponent's perception of your likelihood to follow through. If your threat is too grandiose, your opponent may not take it seriously. To address this issue, you can employ the following strategies:

1. **Escalate Gradually**: Start with a minor threat and increase it progressively to a larger one.

2. **Enhance Believability**: Attach a probability to the risk to make a big threat more credible. This tactic can be achieved by incorporating one or more of the following elements:

   a. *Random Element*: Introduce unpredictability, such as threatening to blow up a building without specifying which one. This addition expands the threat to all buildings, making it more believable.

   b. *Selective Element*: Focus on a particular

target while implying broader consequences. For instance, a terrorist with a gun trained on hostages can threaten to shoot one, thereby expanding the threat to all hostages.

c. *Accidental Element*: Highlight the potential for unintended consequences. For example, trying to free hostages from a terrorist wearing explosives. Even an accidental shot could trigger the explosives, expanding the threat to all hostages.

By carefully managing the size and presentation of your threat, you can increase its credibility and influence your opponent's decision-making.

Tipping the Scales

This strategy is for bystanders and opportunists. First, you must recognize when tensions are explosively converging. Then, you need to trigger the spark necessary to set off that explosion. For example, a slight snicker can ignite an outburst of uncontrollable laughter in a nervous crowd. In the 1950s, unscrupulous real estate agents exploited this tactic by creating panic selling. They would sell one house to a minority, then capitalize on the surrounding neighbors' prejudice and fear, profiting from the resulting panic selling.

Torture

Torture is often considered an effective strategy in the absence of laws prohibiting its use. To be clear, we do not advocate for torture or lawless societies. However, throughout history, torture has played a significant role in autocratic regimes. It has frequently been used to enforce behavior and extract information. Often, torture is more mental than physical, as mental torture is typically less obvious and more tolerated by society.

One primary objective of torture is to extract information, a crucial element in executing any strategy. In this way, torture is used as a means to further a larger strategic goal by gathering necessary intelligence. Another objective of torture is to enforce behavior, often serving as a warning to others rather than just a punishment for the individual subjected to it. This use of torture supports a broader strategy by deterring undesirable actions through fear and intimidation.

Trojan Horse

This strategy involves placing a seemingly harmless structure, organization, or individual at the heart of your opponent's center of gravity—their core strength. Beneath this innocuous exterior lies a sinister interior. Hidden within might be ulterior motives, secret plans, or concealed strengths that emerge under the cover of darkness.

A recent example involves allegations of corporate espionage, such as those involving Chinese technology firms. For instance, in 2020, the U.S.

Department of Justice charged two Chinese nationals with hacking into the systems of companies and organizations engaged in COVID-19 research. These hackers posed as harmless entities but gathered sensitive information to benefit their government.

Turn the Tables

In this strategy, you take an attack and turn it around on your opponent, using their actions against them. To achieve this, you typically need inside information on your opponent, giving you the leverage to reverse the situation. This inside information allows you to anticipate their moves, exploit their weaknesses, and use their strategies against them. By turning their attack back on them, you neutralize the threat and put your opponent on the defensive, gaining a strategic advantage. This approach requires careful planning, precise timing, and a deep understanding of your opponent's tactics and vulnerabilities.

Unreasonable Demand

Using an unreasonable demand can be an effective tactic to unsettle your opponent. To counter this strategy, you can initially request that the unreasonable demand be submitted in writing. This request tactic serves multiple purposes: it provides additional time to formulate a response, often leads to a softening of the original demand, and brings the demand into public view, which can further temper its impact.

Furthermore, you can request additional research or supporting data before responding to the demand. This extra step not only buys you more time but also places the onus on your opponent to substantiate their unreasonable request, potentially exposing any weaknesses or impracticalities inherent in their position. By employing these tactics, you create a more controlled environment to address the demand, diminishing its disruptive potential and allowing for a more strategic and measured response.

Using the Law

Laws are the official rules of the game. Understanding and using these rules to your advantage can be a powerful strategy. For example, you might file a lawsuit against your opponent and generate publicity around the alleged charges. Even if the claims are unfounded and eventually dismissed, the initial announcement often garners more attention and news coverage than the later resolution of the legal proceedings. This tactic can damage your opponent's reputation and create a strategic advantage for you, regardless of the lawsuit's outcome.

A current example of this strategy can be seen in the high-profile legal battles involving major tech companies. In 2021, Epic Games sued Apple for alleged App Store antitrust violations. Epic's initial filing and public statements generated significant media coverage, casting Apple in a negative light. Even though the case's outcomes are complex and ongoing, the initial lawsuit brought considerable attention to Apple's business practices. It sparked widespread debate, highlighting how legal actions can strategically influence public perception.

Verbal Retort

You can use a witty remark as an attack on your opponent. In response, your opponent might employ a counterstrategy, ranging from simply stating that they are ignoring you to launching a more aggressive personal attack. Such a retort typically provokes retaliation, either immediately or in the near future. It may be useful to get a "rise" out of your opponent to show their volatility.

Vulnerability of Importance

Former Mexican President Salinas once remarked that the more important you are, the more vulnerable you become. This vulnerability extends not only to criticism of your programs but also to your security team.

Waiting Game

When you employ the waiting game, you leverage time to your advantage. By stalling or delaying action, you can maneuver your opponent into a more favorable position for your attack, deplete their resources, or otherwise strengthen your position.

Walking the Plank

This strategy originates from the days of buccaneers when people were forced to walk off the plank of a ship into the ocean. It has two distinct versions.

The first version involves being compelled to do something against your will. For example, you might have to pay a fine due to an employee's illegal actions.

The second version of the strategy is more practical, involving the choice of one action to avoid a worse outcome. For instance, E.F. Hutton once agreed to a sweeping reorganization of its legal and financial departments and to strengthen its auditing procedures. This was a proactive step to avoid much more severe penalties for their employees' involvement in a check-kiting scheme.

War

A war is a full-scale attack employing all available resources to achieve victory over an opponent. The objectives of such an endeavor are to:

- Ultimately, control your opponent entirely.

- Seize control of your opponent's territory.

- If possible, eradicate the opponent.

- Monopolize the market by dominating a segment, eliminating competition, and controlling pricing and supply.

- Acquire competitors through mergers and

acquisitions, consolidating market power.

- Undermine opponents using aggressive marketing, pricing strategies, or legal action to weaken their positions.

- Secure resources, such as raw materials, technology, or talent, to outmaneuver competitors.

- Establish brand supremacy, creating a powerful and reputable brand that becomes the preferred choice, overshadowing all rivals.

War also manifests in the business world. In this context, competition, innovation, and superior productivity are tools of corporate warfare. The ultimate goal in military and business wars is to emerge as the dominant force, ensuring long-term control and success.

War of Attrition

A prolonged war aimed at wearing down the opposition's resources, spirit, or will to continue is known as a war of attrition. This strategy is most effective when opponents are of equal strength or when one believes the other will mentally or physically tire from the continuous struggle.

This strategy is particularly effective when your opponent is defending territory that is not their homeland. In a war of attrition, your opponent's lack of emotional attachment to the land becomes a vulnerability you can exploit. Their diminished morale and commitment to the fight can be leveraged to your advantage, gradually eroding their resolve and resources until they can no longer sustain the conflict, showcasing the effectiveness of this strategy.

Wing Wang (Whipsaw)

With this strategy, you act unpredictably by "bending with the wind." You handle situations one way at one time and differently the next. The degree of your randomness or inconsistency confuses your opponent, throwing them off balance as they try to anticipate your actions. This tactic, often called the "old wing-wang," keeps your opponent guessing and unsettled.

A variation of this strategy is the "whipsaw," where you repeatedly change your stance on the same topic to frustrate your opponent further. This approach can be helpful in offensive and defensive scenarios, making it difficult for your opponent to form a coherent counter-strategy.

"Wounded Chicken"

Chickens will peck to death any chicken that becomes sick or wounded. Similarly, in a political environment, people will ruthlessly target someone who becomes politically vulnerable. The challenge is to politically wound an opponent without it being evident that you were responsible. Once your opponent is injured, it doesn't matter if your involvement becomes known.

An illustrative example comes from my

brother Fred's childhood in Catholic grade school. One day, he proudly wore a new red sweater to class, only to be vehemently rebuked by a nun who demanded he remove it, saying it reminded her of Satan. The next day, feeling rebellious, he wore a slightly darker red sweater and was again publicly chastised and ordered to remove it. Not wanting to push his luck a third time, he chose not to wear a sweater the following day. To his surprise, his classmates, led by someone not his friend, all wore red sweaters in a coordinated act of rebellion. When the nun arrived and saw the blatant defiance, she assumed Fred was the ringleader and sent him to the principal's office, resulting in a three-day suspension. Fred had difficulty explaining to our parents why he got suspended for not wearing a red sweater. The instigator cleverly exploited the situation to set up my brother as the "wounded chicken," ensuring he bore the brunt of the punishment.

Recap

The offensive strategies described above often result in limited victories rather than final triumphs. As we will discuss later, achieving total victory requires the continuous application of resources and various strategies against your opponent until your main objective is attained. Additionally, intermediate gains can be diminished or wiped out by chance, political, environmental, or economic changes. Therefore, it is crucial to persist in your strategy to retain these gains. A potent offense also serves as a robust defense, helping to secure the advances you've made as long as the offensive efforts continue.

Offenses - Conclusions

We conclude this section with some summary conclusions about the offensive strategies discussed above:

1. **Follow the Basic Rules**

   - *Focus*: A sharp focus on your objective is crucial after establishing essential strength.

   - *Conserve Resources*: Use only the necessary amount to accomplish the task at hand, nothing more.

   - *Maintain the Right Distance*: Avoid long-range combat, ambushes, or well-organized firing lines.

   - *Use the Right Tool for the Job*: Deploy the appropriate weapon in the right circumstance for optimal results.

2. **Fight Fire with Fire**

   - Match your strength against your opponent's strength.

3. **Sequence Your Attack**

   - Attack your opponent's components of strength in the correct sequence. Neutralize long-range artillery before engaging their troops.

4. **Find the Niche**

   - Target the weakest spot in a restricted part of the total field of strength. Avoid trying to attack everywhere simultaneously.

5. Soften Them Up

- Before launching a full-scale attack, strike at some limited strategic points.

6. Wear Them Down

- Use a combination of long-range attacks, short-range engagements, and close combat to weaken your opponent progressively.

7. Proceed at the Quickest Pace

- Advance as swiftly as possible with reasonable regard for safety.

8. Know Your Enemy

- Understand your objective, your means, those of your opponent, and the character of your troops and those opposing you.

9. Turn Your Enemy: The correct action will cause your enemy to turn and flee at a certain point in the offense. Follow these general rules:

- *Decisive Superiority*: Attempt to turn your enemy only when you have a clear superiority over them.

- *Unified Forces*: Keep your forces together and direct your action toward the heart of your enemy. Avoid attacking while moving parallel, as you are equally exposed.

- *Avoid Surrounding*: Do not encircle the enemy, as this can split your forces. Allow them to flee so you can attack them in their disorder and confusion. The most significant losses occur at this point.

- *Tactical Separation*: Crush whatever you can separate from the main body or base.

- *Merciless Pursuit*: When your opponent flees in confusion, pursue them relentlessly and without mercy.

This summary simplifies the lessons to be learned from the entire body of offensive strategies discussed. It only begins to identify the many existing strategies, providing a foundation for further exploration and application.

When to Stop

The time to end your offense is when success (or failure) is complete. Your objective is nothing short of complete victory. Do not stop too soon. Do not leave your opponent wounded and able to fight again. Do not even start unless you are certain of success and capable of dispatching your opponent and his or her resources completely.

12. Defensive Strategies and Tactics

Clausewitz said that a complex strategy requires intelligence, while a simple strategy only requires courage. If the enemy can move quickly and simply, you should match your strengths and defenses to their capabilities and tendencies within the context of the engagement.

Understanding your opponent's propensity for action is crucial in defensive strategies. Consider the behavior of methodical individuals, who consistently use the same forms of offense and defense, making their behavior predictable. In your defense, avoid being methodical. Instead, identify situations where problems can be avoided by inaction, and problems can be created by action.

In essence, defense has three key components: people, perimeter, and time. The primary asset of defense is people, who fill various roles; volunteers and allies are especially valuable. The primary location of defense is the perimeter, including barriers. The main goal is to preserve time, which may require a range of unconventional delay tactics. For example, a defender might place logs on a railroad track or push a captured favorite son to the front as an obvious target to be avoided. Defense means doing whatever is necessary to preserve your strength.

The list of defensive strategies is shorter than the list of offensive strategies, not because defenses are less essential or because offensive strategies are more successful. Rather, it is shorter because many aspects of offense apply equally to defense, and we are not repeating them here. Conversely, elements of defense can also be used offensively. Here are some defensive strategies:

Adaptability

Adaptability in defense means avoiding predictability. Regularly changing your tactics and strategies keeps your opponent guessing and prevents them from exploiting any patterns in your behavior. For example, during World War II, the British used deceptive tactics, such as fake tanks and misleading radio transmissions, to confuse the Germans about the location of the D-Day invasion. This adaptability prevented the Germans from mounting a successful defense against the actual invasion site.

Absurdity

Absurdity as a strategy introduces confusion to disrupt your opponent's plans. By acting irrationally or unpredictably, you create uncertainty and force your opponent to question and reassess their strategies. This tactic is particularly effective against opponents who overanalyze and overstrategize, often losing sight of their primary objectives. For example, during negotiations, making an unexpected and seemingly nonsensical demand can throw your opponent off balance, causing them to reconsider their approach and potentially make mistakes. This tactic leverages the power of unpredictability to weaken your opponent's strategic coherence and decision-making process.

Against Coalitions

People form coalitions with their friends and allies, creating a network of minor and major alliances. When your opponent is part of a strong coalition, attacking them becomes nearly impossible, and the consequences of attempting to do so can be disastrous. If a member of a strong coalition chooses to attack you, defending yourself can be equally challenging. Your options include:

1. **Creating Your Coalition**: Forming your alliance can bolster your strength, though this is often impractical within the necessary timeframe.

2. **Seeking Higher Authority**: While seeking aid or counsel from a higher organizational element than the coalition can be beneficial, it is not without its risks. It's essential to weigh the potential benefits against the possible consequences.

3. **Retreating:** Abandoning the object of attack and fleeing to avoid confrontation with a powerful coalition.

Alliances

Leveraging volunteers and allies can significantly bolster your defensive efforts. Forming alliances and securing support from others increases your strength and resources. During the American Revolutionary War, the colonies formed an alliance with France, which provided crucial military and financial support, helping to turn the tide against the British.

Arbitration

Arbitration is a process that involves the use of a neutral third party to whom both sides agree to delegate all decision-making responsibility. They also agree to be bound by the resulting decision. This strategy is effective for avoiding total conflict, such as a labor strike or major economic loss, and prevents you from conceding entirely to your opponent's demands. However, be prepared for some of their demands to be fulfilled. Arbitration is useful if you are willing to end up roughly halfway between your position and your opponent's.

Labor negotiations, often viewed as extortion, have led to many improvements in working conditions. Arbitration, though a partial surrender to terms dictated by others, is not ideal for purely economic issues like pay rates. However, it can be beneficial for addressing structural, complex work rule issues with low financial impact.

In arbitration, you forfeit some future

alternatives and your freedom of choice. Binding arbitration, where a third party makes final decisions, can result in permanent changes that typically cannot be reversed. This approach is most suitable when the "boundary" of acceptable solutions is well-defined, and the potential outcomes are acceptable to you. Your decision to accept arbitration should be based on a firm conclusion that it will benefit you more or hurt you less than settlements made without arbitration.

Attack versus Defend

Defense can ideally transition into offense. Deciding whether to be on defense or offense should be based on the following advantages:

1. **Surprise**: If a quick reversal of roles can give you an immediate and significant advantage, seize the opportunity and go on the offensive. Catching your opponent off guard can lead to a swift victory.

2. **Numerical Superiority**: Adjust your strategy based on relative strength. If your strength is relative to your opponent's increases due to reinforcement, weakening of your opponent, or improved morale, go on the offensive. Conversely, switch to defense if your strength diminishes due to attrition, dispersion, or other factors. The strong party should always attack, while the weaker party should defend.

3. **Benefit of Terrain/Environment**: Take advantage of the locale or situation to change

roles. For example, if you are attacking and your opponent begins a counterattack in a terrain that favors defense, switch to defense to rest your troops and consolidate your position. Conversely, if you have been defending and are driven from advantageous terrain, a sudden counterattack might regain the position.

4. **Concentric Envelopment**: If you have sufficiently encircled your opponent to leverage crossfire, attacking is the best action. Conversely, if you find yourself surrounded, a timely attack might break the encirclement and avoid the detrimental effects of crossfire.

5. **Popular Support**: If initiating an attack can garner additional support and strength, making victory more likely, then go on the offensive. Popular support can provide a critical boost to your efforts and sway the momentum in your favor.

Considering these factors, you can effectively decide when to transition from defense to offense, maximizing your strategic advantage and increasing the likelihood of success.

Backup Plan

For every action, you should have at least one backup plan. A backup plan provides a contingency if your initial action doesn't proceed as expected. For example, if you have a secret agent operating in a

high-risk area, you might place another agent within
the local police agency. This second agent can inform
you if your primary agent is captured, allowing you
to either facilitate an escape or take other necessary
actions. In some cases, having multiple layers of
backup plans is advisable, ensuring you are prepared
for various potential outcomes and can adapt swiftly
to changing circumstances.

Bluffing

Bluffing can also serve as a defensive strategy.
As described earlier, bluffing injects risk into your
strategy by creating the illusion of having more
resources than you possess. This trade-off, though
rarely justified, can sometimes be an act of
desperation. For bluffing to work, your opponent
must remain unaware of your desperation. When
considering a bluff, you typically have one of two
motives: to project strength when you are weak or to
project weakness when you are strong. Both motives
aim to mislead your opponent. The first is most
successful when it convinces, and the second when it
deceives.

Bluffing can prevent your opponent from
executing their planned strategy. It includes the
following tactics:

Bluffing Deliberately: Bluff for effect, ideally
at moderate cost. For example, in poker, this would
be like making a modest bet to suggest a strong hand
without risking too much.

Bluffing Strategically: Invest significantly in
bluffing to keep your opponent in check. In poker,

this would be like making a large bet or going all-in to intimidate opponents and force them to fold, even if your hand is weak.

John McDonald describes John von Neumann as distinguishing two varieties of bluffing:

1. **Aggressive Bluffing**: Used when you already have the initiative to strengthen your perceived advantage. In a card game, this might mean continuing to raise the stakes when you have a solid hand, making your opponents believe you are even stronger than you are.

2. **Defensive Bluffing**: Employed to challenge your opponent periodically when you suspect they are exploiting their initiative, effectively "keeping them honest." In poker, this would be like calling or raising occasionally when you suspect an opponent is bluffing, preventing them from becoming too confident in their bluffs.

By understanding and applying these principles, bluffing can be a versatile tool in your strategic arsenal, whether to gain or maintain an advantage. However, there is a risk of bluffing too often. To avoid this, occasionally aggressively bet when you are confident in your results (i.e., having a winning hand after what appeared to be a bluff). This tactic keeps your opponent uncertain and reinforces the credibility of your bluffs.

Box Out

Forming a box formation was a common infantry defense against cavalry forces in the 1800s. This tactic involved arranging soldiers into a square, simultaneously protecting the front, flanks, and rear. A similar defensive tactic was forming a circle, ensuring everyone's back was protected. However, cavalry forces eventually learned to exploit the concentration of soldiers in such formations.

After a missed free throw in basketball, defenders employ a strategic tactic known as 'boxing out.' The players closest to the basket crowd under it to secure the rebound, effectively preventing the opposing team from getting the ball. This strategic move mirrors the protective box formation used in military tactics.

In business, a similar defense occurs when a leader is under attack. Loyal team members will metaphorically "circle the wagons" around the leader, providing protection and support. This collective defense can help shield the leader from external threats and stabilize the situation.

Producing Creativity under Pressure

Creativity can either cease or kick into high gear when in trouble. Experienced leaders often find a way out of problems, while novices freeze up. If you find your creativity stalling, seek help and look for ideas everywhere.

Consider the following strategies:

1. **Remove Bottlenecks**: Identify problems (which may be opportunities in disguise) and determine what bottlenecks prevent a solution. Focus on removing these obstacles.

2. **Create New Combinations**: Evaluate existing combinations and mentally experiment with new ones. If you can't change your overall plan, adjust the delivery method. This tactic could involve changing teams, approaches, resource distribution, or resource application.

3. **Maximize Your Options**: Identify the fundamental needs and explore alternative ways to meet them. Avoid letting others dictate your actions, and don't simply react to match your opponent's moves.

    - *Look for Different Approaches*:

    - *Turn to Your Best Resources*: Utilize your highest quality, best, brightest, and bravest assets.

    - *Maximize Impact with Minimal Resources*: Focus on where you can have the most impact with the least expenditure.

    - *Seek New Forms of Protection*: Innovate ways to safeguard your interests.

    - *Leverage Your Expertise*: Use your know-how and what you do best.

    - *Innovate*: Develop entirely new

approaches.

- *Identify New Capacity*: Look for untapped potential and resources.

- *Redirect Resources*: Drop what isn't working and redirect those resources to new, promising avenues.

Evaluate all your actions against your opponent. Assess what has been working, what hasn't, and what new strategies you could implement. You can navigate through trouble and find practical solutions by staying flexible and open to new ideas.

Common Peril and Binding

Do you have dissension among your troops? Even one person working against your interests can undermine and ruin the efforts of many. If dissenters are not promptly separated, your progress will be impaired, and your success will be threatened. A strong defense against dissension is to bind the group to you as tightly as possible.

Binding the group is crucial; the "fear of hanging separately" will encourage them to "hang together." One effective method to achieve this is by exposing them to a common peril or taking them through a shared ritual, like hazing. A strong bond is formed when a group must work together to avoid a threat. Fraternities and sororities are built on this principle. A common bond is created by requiring pledges to endure hardships together, such as hazing and hell week. This binding ensures the members remain close-knit and protect each other for years.

By fostering a sense of unity and shared experience, you can mitigate dissension and strengthen the cohesion of your team. This not only resolves immediate issues but also sets a strong foundation for the future, ensuring a more harmonious and productive team.

Communication, Internal

One way to ensure the defensibility of an internal communication is to review your draft with those affected in advance. Although this approach gives them time to prepare a defense, it also offers several benefits:

1. **Ensures the Solidity of Your Case**: Gathering feedback can help you identify and address potential weaknesses, making your argument more robust.

2. **Demonstrates Fairness and Friendliness**: Giving stakeholders fair warning and a chance to influence the content shows that you are considerate and open to input.

3. **Enhances Your Reputation**: To the eventual reader, this process portrays you as a fair, thorough, and diligent person who values collaboration and due process.

Compromise

When faced with a particularly strong defense,

compromise may be the best option for the attacker. Effective compromise relies on positioning. The party that is better prepared and has analyzed the alternatives beforehand will succeed more. The person who makes the first offer often holds the best position, provided the offer is sufficiently aggressive. For example, when splitting a dessert with my brother, I would insist that he cut it into two portions, and then I would choose the first piece. This tactic forced him to divide the portions as fairly as possible.

Cooling-Off Period

A cooling-off period is an excellent strategy for gaining more time. Establishing this period requires agreement from all parties involved. Additional time provides more opportunities and alternatives and can change probabilities and expectations. Conditions may change, causing today's serious problems to evaporate tomorrow. A cooling-off period allows time to accumulate more resources and allies, recover from battle, and bolster strength. This strategy is common in arbitration.

Counter-Defense

A strong offense must effectively deal with a strong defense. The key is gathering intelligence. Learn what your defensive opponent needs for subsistence, their objectives, and what settlement they might accept. You can then tailor your offense accordingly to exploit these insights.

Crossover

In defense, crossover involves moving an important resource from a location under attack to one that is not. This strategy is frequently used. For example, in liability and bankruptcy cases, individuals often transfer assets to another family member to protect them from seizure. Similarly, managers might charge new expenses to a different, below-budget account to avoid variances that require explanation, even if the expense doesn't match the account's purpose.

Crowd Emotion

Bertrand Russell observed that "Timidity [on the part of many] spawns crowd organization [by a few]." Crowd Emotion is a crucial element of leadership power. The ideal situation involves danger significant enough to inspire bravery without being so terrifying that it induces fear. In defense, controlling crowd emotion on both sides of the conflict is essential, keeping your team's morale high while dampening the offensive team's enthusiasm.

Dare

Creating a dare involves drawing a line and challenging your opponent to cross it. This tactic can provide several advantages, such as causing a slight delay if reinforcements are on the way, creating a distraction that reveals a vulnerability, or setting a trap for your opponent when they cross the line.

Dealing with a Screamer

A screamer uses a public forum to voice arguments loudly to gain an advantage. You cannot concede to a screamer or play fair because they are already not playing fair. Screamers are not interested in rational discussion or resolving issues. Your counterstrategy is quickly and loudly objecting with sound counterarguments to prevent others from accepting their position. If the opposing side has a screamer, appeal to those who appointed them to replace the screamer. If they refuse, you should appoint your screamer and work behind the scenes to engage with someone more rational.

Decoy

This strategy involves redirecting your opponent's attention from weaknesses to strengths. For instance, an athlete with an injured shoulder bandaged the opposite shoulder in a Tae Kwon Do competition. This tactic misled the opponent into targeting the healthy shoulder, protecting the injury. This tactic can effectively divert your opponent's focus from your actual vulnerabilities.

False Weakness

Two forms of false weakness can be used to gain a defensive advantage: the "defensive feint" and the "broken wing" strategy.

Defensive Feint

The defensive feint involves deliberately making yourself appear weak to prompt a stronger ally to come to your rescue against your aggressor opponent. This plan requires the presence of a stronger ally who is inclined to intervene on your behalf. It also necessitates that you and your ally have the combined strength to overcome your opponent.

Broken Wing

The broken wing strategy mimics a mother bird pretending to have a broken wing to draw a predator away from her defenseless offspring. By appearing weak, you can manipulate your opponent's focus and actions.

Advantages of Appearing Weak

1. **Divert Attention**: Shift your opponent's focus away from you to stronger targets.

2. **Buy Time**: Gain time to regroup or prepare when you are the sole focus.

3. **Draw Out the Opponent**: Lure your opponent in close for a counterattack or to expose their vulnerabilities.

4. **Reveal Opponent's Strength**: Cause your opponent to "show their hand" and reveal their true strength sooner than they wanted.

If your opponent's primary goal is to neutralize your strength, your appearance of weakness may satisfy their objective, causing them to pause or even stop their aggression.

By strategically employing false weakness, you can create opportunities to gain the upper hand, whether through drawing in support, buying time, or setting up a counterattack.

Deflection

This strategy involves shifting criticism to another person or refocusing concerns on a new problem to divert attention from the original issue. It is advantageous when you are under attack and have no effective defense. For instance, if you are called out for an infraction, you might report a recent good deed or highlight a new, more severe problem. Deflection is highly effective and crucial in the world of big business.

Deflection can also entail leveraging your opponent's natural motion against them. In motion, there is strength and power, which can be redirected to your advantage. Just as judo experts use their opponent's momentum to defend themselves, you can deflect an attack so that its force causes the attacker to miss you. The time your opponent needs to recover allows you to prepare for the next move.

An illustrative example of deflection comes from British Concorde Ltd., one of the manufacturers of the now-retired Concorde airliner. In 1978, the Federal Aviation Administration (FAA) charged Concorde with hydraulic problems, a critical issue for flight control. British Concorde Ltd. responded with a statement saying, "The supersonic airliner has passed safety tests as demanding as the current examination by the Federal Aviation Administration." This statement implied that the Concorde had met all of the FAA's safety standards, deflecting the attack back to the FAA and forcing them to defend the stringency of their tests. However, a closer reading reveals that the statement didn't directly address the FAA's specific concerns but diverted the focus away from the alleged problem.

Delaying Tactics

Delay tactics involve using creative methods to buy time and disrupt your opponent's advance. These tactics include placing obstacles, conducting hit-and-run attacks, or utilizing guerrilla warfare. For example, during the Vietnam War, the Viet Cong used intricate tunnel systems and booby traps to delay and frustrate American forces, effectively prolonging the conflict and draining U.S. resources. Such tactics can force opponents to expend more resources and time, weakening their position.

Denial/Lying

Politicians and corporate leaders often use denial or lying to protect their interests. For example, when the French were accused of sinking a Greenpeace ship in 1985, President François Mitterrand denied government involvement. Despite later revelations that he had lied, his image in France

remained largely untarnished, as the French
accepted such lies for "reasons of state." Politicians
frequently face little to no penalty for lying, as
illustrated when George H.W. Bush declared, "Read
my lips, no new taxes," or when Bill Clinton stated, "I
did not have sexual relations with that woman." Such
statements are often dismissed as part of political
rhetoric, allowing politicians to lie with minimal
repercussions.

Deterrence

In "The Strategy of Conflict," Thomas
Schelling clearly defined a theory for dealing with
threats. Deterrence is a defensive threat aimed at
influencing another person's decision by setting their
expectations. To effectively deter, you must present
evidence that your actions will depend on your
opponent's behavior. You can create a balance that
prevents action by linking your behavior to theirs.
For instance, "If you curse me, then I'll punch you in
the nose" sets a clear expectation: if the first action
doesn't occur, neither will the second. For deterrence
to be successful, threats must be carried out if
challenged; if your opponent curses you, you must
follow through with the punch.

Distance/Movement

Maintaining distance from your opponent is
one way to defend against resource loss. If your
opponent cannot reach you, they cannot harm you.
Like a boxer who is light on his feet, you can avoid

damage by keeping your distance and moving quickly. Movement can also force your opponent to stretch their resources, making them more vulnerable. You may cut their forces in half by looping back on a strung-out opponent, though this tactic risks exposing yourself to a pincer movement.

Escape Ambush

This strategy is used when you anticipate being pursued after an action. Set up an ambush along your retreat path to attack pursuers. The ambush is most effective if your pursuer believes you are in a panic and is gaining on you: fake confusion and misdirection to enhance this illusion. Ensure the size of your ambush force is adequate for the attackers' numbers, or be prepared for your forces to sacrifice themselves to cover your escape. Stretching out the pursuers before the ambush will increase its effectiveness.

Expectation Setting

A key to successful negotiation, arbitration, or mediation is changing the expectations of the involved parties. Consider the Palestine Liberation Organization (PLO). Initially dismissed as insignificant, the PLO is now recognized as a critical player in peace negotiations, with their "rights" deserving consideration. They elevated expectations to a new status quo, solidifying any negotiated settlement. This perception shift created a new baseline from which lasting gains can be secured.

Firing

Dismissing a subordinate from a responsible position can be a crucial defensive action. The primary reasons for firing someone include:

- **Eliminating a Threat**: Removing a potential contender for your position or someone who jeopardizes the success of your endeavors.

- **Protecting a Plan**: Ensuring that your strategic initiatives remain intact and unchallenged.

- **Redirecting Failures**: Shifting blame for failures away from yourself or the core team.

- **Enacting Retribution**: Delivering consequences for previous disloyalties or wrongdoings.

Firing someone is a form of protection, similar to other defensive actions. It typically safeguards your political standing within the organization and ensures the stability of your role and objectives.

Fixed Defenses / Trenching In

"Fixed defenses are futile—the only sure defense is offense, and the efficiency of offense depends on the warlike souls of those conducting it," according to General George S. Patton.

Flexibility

Flexibility in defense means being prepared to shift from defensive to offensive strategies when the situation demands. This dynamic approach allows you to exploit opportunities and counteract your opponent's actions effectively. For example, in the Battle of Stalingrad, the Soviet Red Army transitioned from a defensive stance to a counteroffensive, encircling and defeating the German Sixth Army, thereby turning the tide of the Eastern Front in World War II.

Flop

The news media and opponents seek the flop when a candidate for a major political office begins their campaign. It includes hidden flaws, spoken gaffes, controversial opinions, past mistakes, or any other factors that can be used to embarrass, discredit, or undermine the candidate. To defend against this, candidates should proactively address any potential issues. Candidates can humanize themselves and mitigate damage by bringing up past mistakes, explaining, apologizing, or otherwise excusing them early. Admitting issues after exposure suggests hiding something or hoping not to get caught. Ideally, maintaining a clean record prevents any flops from emerging in the first place.

Fortification

Fortification involves strengthening your perimeter with physical barriers and strategic positioning to create a formidable defense. Building walls, trenches, and other fortifications can significantly slow down or halt an advancing enemy. For instance, the Great Wall of China was constructed to protect against invasions and raids from nomadic tribes, fortifying ancient China's northern borders for centuries.

Foundation Web

In the early stages of many chess matches, moves are made with maximum protection of each piece, creating a foundation of defense where each piece protects the others. This method of building defensive protection allows for swift retaliation when attacked. However, once the web of inter-protection is stretched to its maximum range, further stretching can weaken the defense, potentially cutting off resources and collapsing the entire structure. Therefore, it is crucial to expand the web to a sustainable point and then regroup or shift into an independent offense.

Hit-and-Run

Hit-and-run tactics involve rushing in to strike, maximizing the element of surprise, and then retreating quickly to a place of greater safety. This strategy is more effective against a larger, slower opponent than against a nimble one.

Hunker Down

Hunkering down involves taking the lowest profile possible until a threat passes, minimizing exposure and risk. For example, pheasants often stay hidden and let a hunter walk over them rather than break cover. This strategy can sometimes be turned into an offense. In a 1950s movie, a band of American Indians buried themselves in shallow holes, covered with blankets and dirt, and waited until a troop of cavalry marched into the center. At the right moment, the Indians rose and annihilated the soldiers, turning their defensive hunkering into a surprise attack.

Information

Information is a critical part of communication and plays a crucial role in both defense and offense. Defensively, controlling information is similar to hunkering down; it involves withholding information until absolutely necessary. Always tell the truth, but do not disclose more than is required. Giving away too much information can reveal your objectives, provide your opponent with ideas, and potentially compromise your position. When trading information, ensure that you gain more than you give away.

Killing

While previously described as an offensive strategy for retaliation, killing can also be a defensive measure for self-protection or self-defense. It involves neutralizing a threat to ensure your safety and the safety of your assets or interests.

By incorporating these improved strategies, you can enhance your defensive and offensive capabilities, better manage risks, and respond effectively to various challenges.

Mediation

Mediation is the use of an influencer and communicator to help manage a negotiation. Mediator techniques are to achieve the following:

- Constrain communications
- Influence expectations
- Suggest alternative views, methods, and solutions
- Compare offers and counteroffers without revealing them
- Compile results without revealing detail
- Measure progress
- Authenticate "evidence" provided by either side

The mediator is helpful only when a mutual incentive exists for finding a compromise. In mediation, compromise can become unlikely when either party is entrenched and willing to stand their position regardless of consequence, as in a strike over principle or one party is dealing from a position of strength.

Mediation is a good way to stall for time or to delay an outcome. However, you will often find that bringing a third-party viewpoint into a conflict will result in an unexpected basis for resolving the conflict. You must know what principles are at stake and not compromise those which are core to your beliefs.

No

The anti-drug slogan "Just Say No" can also serve as an effective strategy against various unwanted actions. Saying no can sometimes lead to conflict, such as when you tell your teenager they cannot have the car on Saturday night. However, the ability to say no is crucial in managing personal boundaries and maintaining control over your time and resources.

Many people find it incredibly difficult to say no, leading to significant stress when they take on responsibilities they later regret. By learning to say no, you can prioritize your needs and goals, reducing unnecessary stress and improving overall well-being. This strategic use of no helps you to focus on what truly matters and avoid being overwhelmed by external demands.

Saying no effectively involves clarity and firmness. It requires understanding your limits and being honest about them. Additionally, it's essential to communicate your decision respectfully to maintain positive relationships. Over time, mastering the art of saying no can enhance your ability to make strategic decisions, empowering you to lead a more balanced and fulfilling life.

Overrun

Even the best defenses typically have a weak side. This weak side is the direction where the advantages of terrain and other barriers offer less protection than the primary defenses. It is crucial to remain vigilant and guard against potential attacks from this vulnerable area to avoid being overrun. Recognizing and reinforcing your weak side can prevent adversaries from exploiting it and ensure a more robust defense overall.

Package Trade—Reciprocity

The package trade, a strategy that offers multiple benefits, involves two distinct aspects: one where both you and your opponent mutually agree upon the trade and another where one party acts against the other as a form of payback. The mutually agreed-upon package trade is an effective strategy for resolving multiple key issues simultaneously. By bundling several critical matters together, you can negotiate a comprehensive agreement with your opponent, yielding various side benefits.

For instance, consider a prisoner trade. This exchange reduces the burden of maintaining prisoners and eliminates the compound as a potential target for attack, takes time to accomplish, and projects an image of humanitarianism. Several secondary advantages may arise from this action. The opponent's attitude towards you might become friendlier, potentially paving the way for peace initiatives. Even if peace isn't immediately achieved, the future behavior of your opponent's troops may change—they might be more inclined to surrender, knowing that early release through trade is a possibility, making surrender a more attractive option than death.

In the business world, if you are negotiating with another firm where you are both a buyer and a seller, consider combining these transactions into a single package trade. This strategy can streamline negotiations, foster a collaborative atmosphere, and potentially lead to more favorable terms for both parties.

Package Trade: Tit-for-Tat Retaliation

Retaliation, or package trade, involves trading one damage or punishment for one received earlier to get even. This strategy mirrors the eye-for-an-eye approach, reflecting a basic element of human nature. If you're contemplating revenge, think carefully. Emotions often cloud judgment, leading to poorly conceived plans. However, retaliation can also serve as a counter-defense. It's a strategic move to draw an opponent out of a defensive stance. Look for something to draw them out, such as historical tactics like ransoms, hostages, and the destruction of highly valued treasures before their eyes.

Patience

Patience is an effective strategy against a slow, low-damaging offense. Maintaining patience can be your greatest asset when your opponent's actions cause minimal harm over a prolonged period. By not overreacting to minor provocations, you conserve your energy and resources, allowing you to respond more effectively when it truly matters.

Patience allows you to observe and analyze your opponent's tactics, identify patterns, and develop a deeper understanding of their strategy. This understanding can lead to more calculated and decisive actions when the opportunity arises. Moreover, patience can frustrate your opponent, potentially causing them to make mistakes out of impatience or desperation.

Historically, patience has been a critical element in many successful defensive strategies. For example, during sieges, defenders often relied on their fortifications and supplies, waiting out the attackers who would eventually exhaust their resources. In negotiations, remaining patient can lead to better terms as the other party may reveal their true intentions or weaknesses over time.

Perseverance

This strategy describes the ability to hold on as a crucial defense strategy. Defenders have often succeeded against great odds by outlasting their attackers. One notable example is Vietnam, a country with a fraction of the wealth and power of the United States, which managed to outlast a superpower with some assistance from China. Wars of attrition are a testament to the power of perseverance, demonstrating that endurance and resilience can be as decisive as military strength. Holding on can wear down an opponent's resources, morale, and resolve, ultimately leading to victory despite overwhelming odds.

Pinning the Source

Another effective defense is to deploy a smaller unit to attack the opponent's homeland or main resource base. This tactic distracts your opponent and diverts resources away from their offensive efforts. Such an attack can disrupt supply lines, significantly threatening their resupply capabilities and causing potential losses. When faced with this new threat, the opponent will be forced to redirect considerable attention and resources to defend their supply lines. This shift can cause the opponent to regroup and retreat to their homeland. Using a small offensive force to put the opponent on the defensive can be a highly effective defensive strategy.

Pressure

For every pressure, there is an equal and opposite reaction. If you pressure someone, they will push back with equal force. If you attack someone, they will retaliate. Therefore, a more effective approach is to avoid directly pressing them. Instead, move gradually and smoothly towards your objective with subtle, indirect actions. By disguising the source of your attacks, you can advance without provoking immediate retaliation. This strategy allows you to achieve your goals while minimizing the risk of confrontation and escalation. Such a measured approach can be more sustainable and less likely to trigger a strong defensive response, enabling you to navigate challenges and obstacles more effectively.

Psychological Warfare and PSYOPS

Psychological warfare uses deception, misinformation, and psychological tactics to weaken your opponent's resolve. By creating doubt and fear, you can undermine their morale and willingness to fight. This strategy can be more powerful than physical confrontation, as it targets the enemy's mind and spirit, creating confusion and hesitation.

Operation Fortitude, a standout instance of psychological warfare during World War II, demonstrated its effectiveness. The Allies orchestrated an intricate deception to mislead the Germans about the true location of the D-Day invasion. They employed fake equipment, double agents, and false radio transmissions to create the illusion that the invasion would occur at Pas de

Calais rather than Normandy. This misinformation led the Germans to misallocate their forces, significantly diminishing the effectiveness of their defense when the actual invasion occurred.

Psychological Operations (PSYOPS) are a vital component of psychological warfare. PSYOPS involves planned operations to convey selected information and indicators to audiences to influence their emotions, motives, and objective reasoning. PSYOPS can include propaganda, leaflets, broadcasts, and digital media campaigns. The goal of PSYOPS is to alter the target audience's behavior, which could be enemy forces, civilian populations, or even neutral parties. By continually feeding false information and creating an atmosphere of uncertainty, you can erode the opponent's confidence and ability to respond effectively.

Reconnaissance

Continuous intelligence gathering is essential for anticipating your opponent's moves and adjusting your defense accordingly. Reconnaissance provides valuable information about enemy plans and movements, enabling you to make informed strategic decisions.

A prime example of the importance of intelligence gathering is the Battle of Midway during World War II. U.S. naval intelligence successfully broke Japanese codes and uncovered their plans for an attack. Armed with this crucial information, the U.S. Navy strategically positioned its forces and prepared for the assault. This advanced knowledge allowed the U.S. to ambush the Japanese fleet, turning the tide of the battle and securing a decisive victory.

Relinquishing the Initiative

One effective form of defense, and a method for transitioning from offense to defense, is to halt your attack. Halting does not mean surrendering but shifting to a mode of resistance rather than continued aggression. This change in strategy allows you to conserve resources, reassess your position, and fortify your defenses. Passive resistance is a mild form of this approach, where you stop active combat but continue to resist the opponent's advances through non-violent means. This tactic can be beneficial when confrontation is no longer advantageous, allowing you to maintain resilience and strategic positioning.

Resource Management

Resource management is crucial for sustaining a prolonged defense. Efficiently conserving and utilizing resources ensures you can maintain your defense over an extended period. A prime example is the Siege of Leningrad during World War II. The Soviet defenders meticulously managed their limited food and fuel supplies, rationing carefully and optimizing usage to survive the prolonged German blockade. This strategic resource management allowed them to endure the harsh conditions, withstand the siege, and ultimately maintain their defense until the blockade was lifted.

Retreat

As a powerful strategy for resource preservation without admitting defeat, retreat should be your primary option when confronted with imminent total destruction. A strategic retreat, executed with precision and speed, allows you to conserve your forces by avoiding annihilation and regrouping for future actions. This measure, though it may seem desperate, is a key to long-term survival and resilience in warfare.

When you initiate a retreat, your opponent will likely attempt to:

- Identify your intended destination.

- Strike at that point to sever your communications and supply lines.

- Stop your forces from retreating or divide them to prevent a cohesive withdrawal.

- Exact bonus casualties: Your resources are at the greatest risk of destruction during retreat.

The direction of your retreat is of utmost importance and depends on your strategic context:

- If cooperating with an ally, retreat towards their position for support.

- Otherwise, retreat towards your base, the heart of your territory, or parallel to the frontiers.

Reasons for retreat may include:

- **Regrouping after a lost battle**: Retreating allows your forces to withdraw from an unfavorable position, reorganize, and reinforce before re-engaging the enemy.

- **Avoiding confrontation with superior forces**: A tactical retreat helps avoid direct engagement with an enemy with overwhelming strength, thereby preserving your resources for a more advantageous opportunity.

- **Aligning with your broader strategic plan**: Sometimes, a retreat is necessary to realign your overall strategy and ensure that your actions remain consistent with long-term objectives.

- **Responding to strategic movements by your opponent**: By retreating, you can adapt to changes in your opponent's strategy, avoiding traps or unfavorable conditions they may be setting.

- **Maintaining a buffer distance between you and your opponent**: A controlled retreat can increase the distance between you and the enemy, providing a buffer zone that can aid defense and observation.

- **Keeping your lines of communication open**: Retreating under control ensures that your communication lines remain intact, allowing for effective coordination and command.

- **Gaining a more favorable location for future battles**: Moving to an area with better defensive or logistical advantages can turn the tide in future engagements.

- **Providing time for supplies to catch up**: A retreat can buy time for your supply lines to reach your forces, ensuring they are well-equipped for subsequent actions.

- **Ensuring or increasing proximity to supplies**: Moving closer to your supply bases ensures that your forces remain well-provisioned, which is crucial for sustaining operations.

- **Preserving morale and reducing casualties**: Retreating can prevent unnecessary losses and maintain troop morale, as soldiers are less likely to be demoralized when they know their leadership is prioritizing their safety.

- **Exploiting enemy overextension**: By retreating, you might lure the enemy into extending their supply lines and forces too thinly, creating opportunities for a counterattack.

- **Gaining time for reinforcements to arrive**: A retreat can delay the enemy long enough for reinforcements to bolster your forces, changing the balance of power in your favor.

- **Forcing the enemy into less favorable terrain**: A strategic withdrawal can lead the enemy into difficult or disadvantageous terrain, where your forces have a better chance of success.

- **Implementing scorched earth tactics**: Retreating while denying resources to the enemy (through scorched earth tactics) can weaken their advance and limit their operational capacity.

- **Reassessing and refining strategy**: A retreat allows you to reassess your situation, gather intelligence, and refine your strategy based on the latest developments.

- **Drawing the enemy into a trap**: A feigned retreat can lure the enemy into a predetermined ambush, turning an apparent withdrawal into a decisive offensive maneuver.

A well-executed retreat has the potential to turn a potentially devastating situation into an opportunity for recovery and counteraction. It can preserve your forces and maintain the strategic initiative, giving you a chance to regroup and plan your next move.

Retreating Circumstances

As you retreat, you may encounter a variety of circumstances:

1. **You may not be pursued**: In some cases, the enemy may choose not to follow, allowing you to retreat unchallenged.

2. **Evasion by a small group**: If your group is small, you can evade pursuit through forced marches, night movements, or stealth tactics.

3. **Coordination for a large group**: If your group is large, it should be able to:

   - *Maintain strong rear guards*: Allocate about one-third of your forces to act as rear guards, regularly rotating them to keep them fresh.

   - *Move cohesively*: Ensure the group moves together in parallel or converging lines to maintain organization and support.

   - *Stay combat-ready*: Be prepared to engage in battle anytime during.

   - *Avoid congestion*: Form up and organize to prevent bottlenecks and confusion.

   - *Rear guard tactics*: Have the rear guard periodically attack the pursuers, then retreat to boost morale and slow the enemy's advance.

   - *Engage pursuing inferior forces*: When faced with an inferior pursuing force, engage and defeat them. However, use intelligence to avoid falling into traps.

Secrecy

Following a policy of constant secrecy can prevent your opponent from knowing your plans and may shield you from public criticism. By keeping your strategies and intentions hidden, you maintain an element of unpredictability that can be advantageous in both military and political contexts.

Secrecy is intrinsically tied to the element of surprise. When your actions catch your opponent off guard, you can potentially gain a significant strategic advantage. For example, during wartime, maintaining operational secrecy can prevent the enemy from preparing adequate defenses or countermeasures, thereby increasing the chances of a successful offensive.

However, secrecy must be managed carefully. The case of President Nixon and the Watergate scandal illustrates the risks associated with failed secrecy. Nixon attempted to use secrecy to conceal the Watergate break-in from the U.S. public. The exposure of the secret led to widespread public outrage and ultimately to his resignation. This example underscores that while secrecy can protect and enhance strategic operations, the failure to maintain it can result in severe consequences, including loss of credibility and trust.

Moreover, secrecy can be a two-edged sword. While it can offer strategic advantages, excessive secrecy can breed internal mistrust and lead to breakdowns in communication. It's vital to strike a balance between the need for secrecy and transparent communication within your team to ensure coordination and morale are not compromised.

Striking a balance between the benefits of secrecy and the risk of mistrust involves limiting sensitive information to those directly involved while maintaining transparent communication about the purpose and importance of the secrecy to build trust within the team.

Silence for Truth-Seeking

More psychological than physical, silence was employed as a defensive tactic in a system board union arbitration case. The hearing officer, tasked with determining the truthfulness of a union member's dismissal, used silence as a strategy. When the member denied an accusation, the officer chose to remain silent. The prolonged silence prompted the union member to speak further, attempting to justify his denial. This reaction, a clear sign of deception, serves as a compelling demonstration of the effectiveness of silence as a strategy for uncovering the truth.

Smoke Screen

Similar to the strategy described under Offenses, this tactic originated in warfare when smoke was used to conceal the movement of troops and ships. In business, a "smoke screen" involves raising unrelated issues to obscure the primary, targeted issue. As a defense, a physical smokescreen can help by distracting opponents and buying time to

address the core problem without immediate
scrutiny.

Source of Courage

Taking a strong stand or mounting a robust
defense against a superior attack requires courage.
The question often arises about how one learns or
acquires the ability to be courageous in such
situations. Courage in defense begins with habit,
experience with self-direction, and familiarity with
the environment—the habit of being decisive fosters
a readiness to act under pressure. Experience with
self-direction, developed through a life of
independence and adventurous behavior, builds the
confidence necessary to face overwhelming odds.
Familiarity with the environment where courage is
needed, such as knowing the battlefield in war, helps
eliminate many elements of fear.

In defense, courage is further bolstered by a
sense of duty and responsibility. Defenders often feel
deeply committed to protecting what they hold dear,
whether it be their homeland, loved ones, or
principles. This sense of duty can drive them to take
bold actions even in the face of daunting threats.
Additionally, strategic planning and preparation can
enhance courage, as being well-prepared for
potential scenarios reduces uncertainty and fear.

Therefore, courage in defense is not just an
innate trait but a skill developed through decisive
habits, self-directed experiences, and deep
familiarity with the environment. It is reinforced by a
strong sense of duty and meticulous preparation,

enabling individuals to stand firm and respond
effectively under pressure.

Subsistence

As a conflict prolongs, the issue of subsistence
becomes increasingly crucial. Essentials such as
food, supplies, ammunition, and fuel are vital during
an extended physical conflict. Similarly, subsistence
may be creative ideas, fresh perspectives, and
sustained energy in a political conflict. The ability to
provide these resources at critical moments can often
be decisive. Defensively, maintaining subsistence
buys you valuable time. However, time quickly
becomes a pressing concern if these resources are
depleted.

Starved Out/Siege

Some defensive positions are resource-
limiting, meaning they sacrifice normal supply
channels to maintain a solid defensive stance. In
such situations, your opponent's recommended
offensive strategy will likely wait and attempt to
starve you out. However, waiting can work to your
advantage, as patience is a critical strength in
defense. You might plan for reinforcements to arrive
suddenly from your rear or flank, turning the tables
on your opponent.

Allies are often sympathetic to beleaguered
defenders, increasing the likelihood of receiving
timely support. Additionally, you can suddenly shift
from defense to offense, catching the enemy off

guard. If you realize your situation has become dire, you might consider a bold escape or a "suicide attack," deciding that it's better to die fighting than to perish while defending.

Straw Men

You can deploy someone or something as a shield to absorb your opponent's blows. This strategy could involve putting an issue in the spotlight to see who attacks it, setting up a cover or front person for a questionable venture, or creating a dummy as a decoy. These tactics either protect you or gather valuable information about your opponent's intentions and strategies.

Stonewall

This defense involves finding a way to stall action on a decision, blocking something your opponent wants or needs to be done. Stonewalling requires some form of power, but even those with small yet critical types of power can successfully stonewall many more powerful individuals. For example, the IT department in many companies is known for subtly delaying essential initiatives to enhance the perceived importance of their role. Similarly, holders of expert information often stonewall to bolster their significance within an organization.

Suicide

Suicide can also be a form of defense, albeit one that primarily signifies a recognition of defeat rather than defeat itself. Its defensive nature lies in its impact on the opponent. For example, suicide serves as a defense when a captured spy takes their own life to prevent being tortured into revealing vital secrets. This act denies the opponent crucial information, protecting their cause even in the face of personal loss.

Takeover Defenses

Business takeover defenses encompass a wide range of strategies designed to prevent or deter hostile takeovers. Common defenses include:

- **Taking on debt, such as by acquiring another company**: This can make the company less attractive to potential acquirers by increasing its liabilities.

- **Selling a significant block of its stock to someone else**: Often known as a "white knight" defense, this tactic involves finding a more friendly party to purchase a large stake in the company.

- **Recapitalizing**: Restructuring the company's debt and equity mixture to make a takeover less appealing or more difficult.

- **Organizing its leveraged buyout (LBO)**: The company's management buys out the shareholders to take the company private, often with substantial debt.

- **Attempting to acquire the acquirer**: This defense, known as a "Pac-Man" defense, involves the target company attempting to turn the tables by purchasing the company that is trying to acquire it.

- **Selling and leasing back undervalued assets**: Selling key assets and leasing them back, thus raising cash, will potentially complicate the financial landscape for the would-be acquirer.

- **Reacquiring its shares**: By buying back its stock, the company can reduce the number of shares available for purchase, making it more difficult for an acquirer to gain control.

- **Issuing poison pills**: Issuing new shares or rights to existing shareholders that make a takeover prohibitively expensive or dilutive for the acquirer.

- **Implementing golden parachutes**: Offering lucrative benefits to top executives if they are terminated following a takeover would make the takeover more costly.

- **Staggering the board of directors**: Structuring the board so that only a fraction of the directors can be replaced at any one time slows down the takeover process.

- **Employing the crown jewel defense**: Selling off the company's most valuable assets will reduce its attractiveness to the acquirer.

Three Circles of Protection

According to the TV show Top Cops (11/08/90), the Secret Service at that time employed three circles of protection around the President of the United States: I believe in 2024 they still use this strategy.

1. The military and local police

2. The uniformed Secret Service police

3. The plainclothes Secret Service agents

The Secret Service's tactics were not static but dynamic. They employed various formations around the President, including the box, the triangle, and the circle. The choice of formation was always based on the specific conditions and threats present at the time.

Throwing Up Obstacles

Placing obstacles in your opponent's path can significantly slow them down and disrupt their momentum. These obstacles can take various forms, such as physical barriers, misinformation, or logistical challenges. Physical obstacles like barricades, trenches, or debris can impede movement and force the opponent to take detours, buy time, and potentially expose vulnerabilities.

Logistical obstacles, such as disrupting supply lines or creating administrative bottlenecks, can hinder the opponent's advance. This strategy might involve cutting off access to essential resources or sabotaging infrastructure in warfare. In business, it could mean creating regulatory or contractual complications that delay the opponent's progress.

By strategically placing these obstacles, you can gain valuable time to regroup, reinforce your defenses, or plan a counterattack. This tactic disrupts the opponent's plans and forces them to allocate additional resources and attention to overcoming these hurdles, weakening their offensive capabilities.

Time

Time is a powerful asset in defense. The primary objective of any defensive strategy is to buy time, gradually wearing down the offensive force. Your forces can renew and regroup as time passes, and their morale will likely improve. Techniques for stalling include creating physical barriers to slow the enemy's advance and engaging in delaying tactics such as skirmishes and hit-and-run attacks to disrupt their progress.

Trenching/Fortification

Trenching involves constructing elaborate fixed networks of physical protection. While we often think of the ditches dug during World War I, trenches encompass everything from foxholes in war to becoming entrenched in a market in business.

Although General Patton demonstrated that trenching and other fortifications could be ineffective, Clausewitz identified two valid applications for entrenchment:

1. **Protecting an area of particular strategic significance**: This is effective as

long as the defenders cannot be overrun or starved out.

2. **Expecting help from reserves soon**: Entrenchment is beneficial when reinforcements are anticipated, allowing the defenders to hold their position until help arrives.

Trust and Balance of Power

The defensive element that must be managed most carefully is trust. Trust forms the foundation for seeking peace. If you are on the defense and your opponent is not steadily attacking, it may be the right time for a peace offering. Conversely, if you are preparing a counterattack, your opponent will be unlikely to trust any peace gesture.

Your opponent's ability to trust you is also crucial in using the balance of power as a defensive strategy. If, as some past U.S. presidents have summarized, "the best way to preserve peace is to be prepared for war," then your opponent may see your preparations as a threat, undermining trust and making peace impossible.

Building trust is essential to avoiding conflict. Recent arms reductions by the Russians, aimed at building trust with the United States, illustrate that trust-building measures can be effective but take time. The key to lasting peace lies in fostering genuine trust between opposing parties.

Weakness

In Power: How to Get It, How to Use It, Michael Korda identifies four basic moves for playing a game of weakness:

1. **Deny having any power**: This allows you to avoid the painful necessity of taking a stand on an issue.

2. **Join forces with your opponents**: Position yourself as a fellow victim of the system to gain their sympathy and alliance.

3. **Counter complaints with protestations of your suffering**: Ideally, you should present your grievances first to set the tone.

4. **Make the other person feel guilty**: Use guilt to undermine their position and assert your own.

These moves can be employed in various ways. In a combative situation, you can leverage the egos of your allies' leaders by allowing them to take on a more significant role. By yielding to their need for prominence, you let them bear the brunt of the conflict and weaken the opponent. Once both sides are exhausted, you can then step in as the dominant force, having conserved your strength while others have expended theirs.

"We're Different"

The "we're different" strategy is a version of deflection. In the late 1980s, limited partnerships were created to buy movies. Earl C. Gottschalk, Jr., a staff reporter for the Wall Street Journal, examined their past performance in the July 12, 1989 issue and deemed it poor. The general partner conceded that most previous partnerships past performance was poor but insisted that their strategies were now different. However, no explanation of these purportedly different strategies was provided.

Wing Wang Defense

As in offensive strategy, Wing Wang means to bend with the wind. This strategy involves appearing to cooperate initially, only to reveal non-cooperation later. By giving the appearance of "going along for the moment," one buys time and delays the need to resist openly. This tactic can be difficult to recognize initially; the only clue may be when someone agrees too readily, especially when their cooperation is unexpected.

Gaining genuine cooperation is also crucial for achieving your objectives. There are a few ways to ensure someone's cooperation:

1. **They want to cooperate**: Genuine cooperation occurs when individuals genuinely favor helping you. This attitude may be due to your admirable abilities and persuasive skills or because working with you aligns with their goals.

2. **They are forced to cooperate**: Sometimes, you may corner someone offensively, compelling them to cooperate. However, this approach typically only works once. The coerced party will be wary, prepared to counter any future attempts to corner them, and may seek retaliation when given the opportunity.

Ensuring genuine cooperation through mutual interest is generally more sustainable and effective than coercion.

Defenses - Conclusions

As you consider how each defense may apply to you, consider how your opponent might use these defenses against you. Reflect on how you might react to each defense if you are on the offense. For example, if you decide to starve out a defender, consider what the defender is doing to counter your strategy. Defensive strategies are fundamentally about managing time and protecting resources. Recognizing this, if you can break through a defense, you may be close to securing a victory. Additionally, anticipate how defenders might exploit time to regroup, seek reinforcements, or launch counterattacks. Understanding these dynamics will enhance your ability to adapt and respond effectively in offensive and defensive situations.

12. RECAP

In conclusion, synthesizing the core principles and strategies discussed throughout this work highlights the importance of adaptability, foresight, and calculated action in achieving long-term success and resilience in competitive and defensive scenarios.

1. **Know the rules but do not hesitate to break them when necessary**: Flexibility and adaptability are crucial in dynamic situations.

2. **Anticipate the environment**: Understand terrain, weather, and other environmental factors that can impact your strategy.

3. **Know your motive**; pick the right objectives: Ensure your goals are clear and aligned with your overall mission.

4. **Know your opponent and yourself**: strengths and failings: Deep knowledge of both sides allows for better planning and exploiting weaknesses.

5. **Use a creative approach**; be organized, but do not signal how you are organized: Innovation keeps your strategy unpredictable while maintaining internal coherence.

6. **Be better trained, more practiced, and more assertive**: Superior preparation often translates to superior performance.

7. **Plan just long enough; anticipate dangers, risks, and consequences**: Over-planning can be as detrimental as under-planning.

8. **Define your sequence of events**; organize the flow of resources and logistics: Ensure smooth execution and resource allocation.

9. **Hold back adequate reserves**: Keep a portion of your forces in reserve to respond to unexpected developments.

10. **Start when the time is right and then move fast**: Timing and speed can be decisive factors.

11. **Lead your people; keep attitudes high**: Strong leadership and morale are essential for maintaining momentum.

12. **Strike with concentration and vigor, engage actively, striking at the center of gravity**: Focused and energetic actions are more likely to succeed.

13. **Use surprise if it can conclude the matter in one blow**: Surprise can disrupt and overwhelm your opponent.

14. **Solve problems quickly as they arise**: Rapid problem-solving prevents minor issues from escalating.

15. **Plan your rest and rest when planned**: Scheduled rest ensures your forces remain effective.

16. **Monitor your progress**: Continuously

assess and adjust your strategy as needed.

17. **Maintain strong communication**: Ensure all team members are informed and aligned with the strategy to avoid confusion and enhance coordination.

18. **Adapt to changing circumstances**: Be ready to modify your strategy based on new information or unexpected changes in the situation.

19. **Leverage technology**: Use technological advancements to gain an edge in intelligence, logistics, and execution.

20. **Build alliances**: Strengthen your position by forming alliances that can provide additional support and resources.

21. **Evaluate and learn from each action**: After every major action, continuously assess what worked and what didn't to improve your strategy.

22. **Upon success, follow your plan to annihilate, dominate, or integrate**: Ensure your victory is complete and sustainable.

23. **Upon failure, retreat and, if possible, live to fight another day**: Preserve your forces for future opportunities.

24. **If not or if you must, die for your cause**: Ultimate commitment can inspire and solidify resolve.

Lastly, do not back off when you are winning unless you must rest your resources. Press your offense until your success is both certain and complete. Relentless pressure can turn a favorable position into a decisive victory.

13. REFERENCES

Barbieri, Susan M. "Manipulation May Win You 'The Mind Game,' but in the End You Lose." The Orlando Sentinel, 4 June 1991, p. E1.

Barnum, Cynthia, and Natasha Wolniansky. "Why Americans Fail at Overseas Negotiations." Management Review, October 1989, pp. 55-57.

Blumenson, Martin. The Patton Papers 1940-1945. Houghton Mifflin Company, 1957.

Bloomfield, L. P. Political Gaming. U.S. Army War College, Carlisle Barracks, PA, November 1959. Mimeo.

Callois, R. Man, Play and Games. Thomas and Hudson, 1962.

Cohen, William A. The Art of the Leader. Prentice Hall, 1990. ISBN 0-13-046657-3.

DeKeijzer, Arne J. "Personal Selling Power." Personal Selling Power, Jan./Feb. 1994, vol. 14, no. 1, pp. 12-22.

Dichter, Ernest. The Naked Manager. Cahners Books, Div. of Cahners Pub. Co., Inc., 1974. ISBN 0-8436-0735-1.

Earle, E.M., editor. Makers of Modern Strategy. 1943.
Ellsburg, Daniel. "The Art of Coercion." Sponsored by the Lowell Institute, Boston, March 1959.

Etzioni, Amitai. "Humble Decision-Making." Harvard Business Review, July-August 1989, pp. 122-126.

Evertt, H. "Recursive Games." In Contributions to the Theory of Games, III, edited by M. Dresher, A.W. Tucker, and P. Wolfe, Annals of Mathematics Studies, no. 39. Princeton University Press, 1957, pp. 47-78.

Friedman, Milton, and L.J. Savage. "A Study on Risk-Taking." Journal of Political Economy, August 1948.

Gallagher, Patricia. "Jury's Still Out on Mediation." The Cincinnati Enquirer, 10 December 1990, p. D1.

Gerhardt, Panda. "Setting the Course." US Airways Inflight Magazine, June 1989, pp. 22-29.
Goffman, Erving. "On Face-Work." Psychiatry: Journal for the Study of Interpersonal Processes, 1955, pp. 218-224.

Hall, Stephen S. J. "Ethics in Hospitality: How to Draw Your Line." Lodging, September 1989, pp. 59-61.

Harkins, Colonel Paul D., annotator. War as I Knew It by George S. Patton, Jr. Pyramid Books, May 1966.

Helmer, O. "Safe: A Strategy and Force Evaluation Game." Rand Corporation, RM-3287-PR, October 1972.

Henderson, M. Allen. How Con Games Work. Citadel Press, 1985. ISBN 0-8065-1014-5.

Henderson, M. Allen. Money for Nothing: Rip-offs, Cons and Swindles. Paladin Press, 1986. ISBN 0-87364-389-5.

Hymowitz, Carol. "Five Main Reasons Why Managers Fail." Wall Street Journal, 2 May 1988, Section 2, Page 21.

Hyche, Jerald. "Upcoming Law Would Protect the Time-share Owner." The Tampa Tribune, 22 July 1989.

Korda, Michael. Power, How to Get It, How to Use It. Random House, 1975.

Kriegel, Robert J. "Do You Risk Enough to Succeed?" Reader's Digest, October 1991, pp. 59-62.

Le Poole, Samfrits. "Negotiating with Clint Eastwood in Brussels." Management Review, October 1989, pp. 58-60.

Lipman, Joanne, and Thomas R. King. "Miller Lite Is Dumping Famous Campaign." Wall Street Journal, 2 July 1991, p. B4.

Lykke Jr., Colonel Arthur F., editor. Military Strategy: Theory and Application. U.S. Army War College, May 1986.

Maccoby, Michael. The Gamesman. Simon and Schuster, 1976. ISBN 0-671-22353-4.

Maslow, Jonathan Evan. "The Anatomy of Defeat." Mainliner (United Airlines Inflight Magazine), June 1980, p.

McDonald, John. Strategy in Poker, Business and War. Illustrated by Robert Osborn. W.W. Norton & Co., Inc., 1950.

McGarvey, Robert. "Confidence Pays." U.S. Airways Inflight Magazine, June 1989.

Milnor, J.W., and L.S. Shapley. "On Games of Survival." In Contributions to the Theory of Games, III, edited by M. Dresher, A.W. Tucker, and P. Wolfe. Annals of Mathematics Studies, no. 39, Princeton University Press, 1957, pp. 15-46.

Mostel, Josh. "In Your Face, Mate." Esquire, November 1987, p. 42.

Murray, H.J.R. A History of Board Games Other Than Chess. Oxford Press, 1952.

Newman, James R. The World of Mathematics. Simon and Schuster, 1956.

Ohmae, Dr. Kenichi. The Mind of the Strategist: Business Planning for Competitive Advantage. Penguin Books Ltd., by arrangement with McGraw-Hill Book Company, 1982. ISBN 0-1400-9128-9.

Owen, Wilfred. "Strategy for Mobility." The Brookings Institute, Transport Research Program, 1964.

Patton, Charles D. Colt Terry, Green Beret. Texas A&M University Press, 2005.

Patton, George S., Jr. War as I Knew It. Annotated by Colonel Paul D. Harkins. Pyramid Books, May 1966. Published with arrangement with Houghton Mifflin, Co.

Paret, Peter, and Michael Howard, editors and translators. On War by Carl von Clausewitz. Princeton University Press, 1976. Originally translated from Vom Kriege by Colonel J.J. Graham in 1874.

Porter, Michael E., and Victor E. Millar. "How Information Gives You Competitive Advantage." Harvard Business Review, vol. 63, no. 4, July-August 1985, pp. 149-160.

Richardson, Lewis Fry. "Mathematics of War and Foreign Politics." See item 16 above, pp. 1240-1253.

Rosefsky, Robert S. Frauds, Swindles and Rackets. Follett Publishing Co., 1973. ISBN 0-695-80384-0.

Russell, Bertrand. Power: A New Social Analysis. W.W. Norton & Co., 1939.

Russell, Bertrand. Authority and the Individual. Simon and Schuster, 1949.

Schnelling, Thomas C. The Strategy of Conflict. Harvard University, 1960, 3rd Printing 1966. LC60-11560, 309 pages.

Schilit, Warren Keith. Review, November 1988, pp. 41-44.

Shapley, L.S. "Stochastic Games." Proceedings of the National Academy of Sciences, vol. 39, 1953, pp. 1095-1100.

Shubik, Martin. Strategy and Market Structure. Wiley, 1959.

Shubik, Martin, and G.L. Thompson. "Games of Economic Survival." Naval Research Logistics Quarterly, vol. 6, no. 2, 1958, pp. 111-123. Also Shapley and Shubik, op. cit., chap 3.

Stern, Walter H. The Game of Office Politics: How to Play It to Win. Henry Regnery Co., Chicago, 1976. ISBN 0-8092-8157-0, 208 pages.

Sun-Tzu. The Art of Warfare. Translated and commented by Roger T. Ames. Ballentine, 1993. ISBN 0-345-36239-X.

Tedeschi, J.T., B.R. Schlenker, and T.V. Bonoma. Conflict, Power, and Games. Aldine, 1973.

Terhune, K.W. "The Effects of Personality in Cooperation and Conflict." In The Structure of Conflict, edited by P. Swingle. Academic Press, 1970, pp. 193-234.

Vajda, S. "Theory of Games." See item 16 above, pp. 1285-1293.

Von Neumann, John, and Oskar Morgenstern. Theory of Games and Economic Behavior. Princeton University Press, 2nd Edition Revised, 1947.

Wessel, Harry. "Scapegoating." Orlando Sentinel, 3 December 2003, p. F1.
"Putin 'to Manipulate Trump with Flattery' as They Come Face to Face for First Time at G20." *The Independent*, 7 July 2017, www.independent.co.uk/news/world/americas/us-politics/putin-trump-g20-flattery-manipulate-sociopathic-narcissist-ex-spies-predicted-a7829431.html. Accessed 1 July 2024.
Waldman, Paul. "Everyone Has Figured Out How Susceptible Trump Is to Flattery—Except Trump." *The Week*, 18 Dec. 2017, www.theweek.com/articles/744807/everyone-figured-how-susceptible-trump-flattery--except-trump. Accessed 1 July 2024.

Zaleznik, Abraham, and Manfred F.R. Kets de Vries. Power and the Corporate Mind. Houghton Mifflin Co., 1975. ISBN 0-395-20426-7, 288 pages.

INDEX

ALSO BY CHARLES PATTON

• **For Honest Citizens Only**

A bold call for Americans to rise above politics and rebuild civic integrity.

• **In Defense of the Righteous**

A gripping story of moral courage when justice and survival collide.

• **Tigers of the Ice**

Adventure meets survival in an unforgiving world where instinct rules.

• **Mastering Strategy**

The essential guide to thinking, planning, and winning in any field.

• **Thinking**

Learn how to think more clearly, act decisively, and change your life.

• **Artificial Consciousness**

Explores the frontier of automating consciousness.

• **The Gardener's Secret and Other Stories**

Mysteries and dramas revealing the hidden motives behind ordinary lives.

• **Extreme Leadership**

What real leaders do when the stakes are high.

• **Who Do You Trust**

A deadly game of deceit between two spies and one truth.

• **Busted, What's Wrong With My Excuse**

An entertaining look at excuses people make, and how to excuse better.

• **Naked Reflections**

Raw, honest poetry of truth, ego, and the search for authenticity.

• **Charles Patton, Visionaire**

Insights from a lifetime of ideas, invention, and fearless creativity.

• **Storming the Castle Bridge**

A tale of rebellion, loyalty, and the unbreakable human will

Find every title at: charlespattonbooks.com

www.ingramcontent.com/pod-product-compliance
Lightning Source LLC
Chambersburg PA
CBHW062112020426
42335CB00013B/940